Surviving the Apocalypse

Surviving the Apocalypse

Jacqueline Druga-Marchetti

Writers Club Press
San Jose New York Lincoln Shanghai

Surviving the Apocalypse

Writers Club Press
an imprint of iUniverse, Inc.

For information address:
iUniverse, Inc.
5220 S. 16th St., Suite 200
Lincoln, NE 68512
www.iuniverse.com

ISBN: 0-595-20693-X

Printed in the United States of America

To my mother, Carol. For setting me on the fearless and inquisitive path of apocalyptic discovery and knowledge that made me the writer I am today.

What will we become?
Ashes in the wind, whispers in the dark, a memory faded in the
blink of an eye.
Strength of our past, weakness of the future.
Animals of our evolution that reappear in the pressure of survival.
Will we still be man?
What will we become?

—J. Druga-Marchettii

Contents

◆

Foreword

---◆---

There are a lot of things they just don't teach our kids in school, perhaps some things just can't be taught. An appreciation for and a respect of the little things that make up the world around us, is something that each of us has to come to recognize and learn to respect in our own way. To me, survival, or more specifically, survival skills, draw upon that respect and depend on one's ability to be observant, logical and level headed in unusual or extraordinary circumstances.

Although I am an American citizen, I've spent the better part of my life living in Western Canada. I've always loved the outdoors and take every opportunity to fully enjoy Canada's Rocky Mountains. I've also been involved with Canada's youth Scouting program for about 15 years. I don't consider myself an expert in survival, but more of a student (I'm always learning). I like to think of myself as a facilitator for youth to get out into the wilderness to discover the many wonders that most of us take for granted. One never knows where or when one might have to use survival skills. Life can at any moment thrust us into a position where we are forced to use nothing but our most basic instincts. Growing up and living most of our lives in concrete, urban jungles, many of us lose touch with our instincts. I feel that getting kids out into the wilderness, gives them an opportunity to develop skills and knowledge that may, some day, help save their lives.

Before I take a group of kids out on a wilderness adventure, I spend some time going over details of the trip. We can't take a refrigerator, so special consideration has to be given to food preparation, packaging and storage. We discuss choice of clothing. Staying warm, even during summer months is much more important than fashion. Other issues such as choice of sleeping bag, how to defecate in the woods and how to avoid becoming a bear's lunch are also discussed. We talk a bit about how to live off the land, but I find it of much greater benefit to the kids to try various edible plants we find along the trail. We show the kids how to set rabbit snares and how to prepare the beast for cooking. Kids love fishing, so we try to fit that in wherever possible.

Survival is more of an attitude. First, you have to have the desire to survive. In order to survive, you must maintain your basic needs. You need air, you must remain warm and dry, you need a source of drinkable water and sooner or later, you will need food. If those first 5 necessities can be met, the rest is attitude. Stay calm, think logically and know how to utilize the surrounding environment to your advantage.

By Michael Lipovski

Preface

◆

November 1963.

I guess you can say that is what started it all. Not long after the Cuban Missile Crisis, worries still predominantly on mind, Vietnam, uprising, fear of Communism.

It happened.

John F. Kennedy was assassinated.

Of course, it had an impact on everyone. It held a great impact on history, the world. But there is one individual I would like to tell you about.

She was a young woman then, twenty-three. President Kennedy's assassination not only impacted her, but changed her life…literally.

Fear set into this woman. With all that had happened in the world, she was convinced that the time was at hand. Surely, mankind was about to meet extinction.

The world was going to end.

The wrath of God was underway.

Very religiously raised, through education and family, this woman was certain, that without a doubt, the signs were all there. The bible was her reference manual. The Book of Revelation held the prophetic listing of all that had happened and that would unfold.

There was no turning it back. It had already begun. Her prayers to the Heavens above were more so for repentance than for anything else. After all, God had made his decision. Her days were filled with neuroticism.

Her nights were filled with tears. Not even the innocent smile of her two-year-old son could break through the veil of destruction and gloom that hung over her head.

It would be classified as nothing other than unconditional fear. It took over her. Then her husband, took her to a doctor.

He was an older physician who pretty much had heard and seen it all. There was no ridicule from him, no prescription for medication, nor recommendation to any type of institute. His advice was quite simply…get pregnant.

He knew no amount of logical talking would convince the young woman the world wasn't ending, but he was certain that carrying a child would take her mind off of it.

She followed his advice.

It worked.

Rather quickly, after finding out she was with child, the woman's fears left her and she soon saw the errors of her thinking. No more fear was with her. To her if it was going to happen, what could she do about it? No longer was she obsessed with the end of the world. That obsession left her. However, somehow, someway, that obsession was passed on to the child.

The woman is my mother. The child is me.

Which brings us to this book.

Since I could remember, I have been obsessed with any and all apocalyptic scenarios. Not laced with fear, but of curiosity. My fictional novels reflect that.

It was through multitudes of research, years' worth, that I have accumulated a virtual vat of information to place into my fictional stories. It was after long contemplation that I realized the information alone, was enough to be a book in itself.

So here we are.

It is not my intention to make this guide a scientific work of wonder to be forever etched in stone. But merely an informational manual with viable and practical knowledge that can help you understand, and survive two different destruction scenarios.

The two scenarios are real. The facts are real. It is my hope, and belief, that what you learn through this book, will be no more than an ounce of knowledge you will carry with you, and never have to use.

Acknowledgements

◆────────

To my son, Drew, for all the interest that you showed in learning about survival. My children for putting up with my relentlessness. Steve, for listening to me ramble. Rob, for always encouraging me. Terri, for your continued support and interest. Monica, you believe in me, and that means more than any words could say. Heather, our talks and arguments made me see that I have a lot to teach and to say. To my father, for teaching me not to fear the inevitable. Mike from Canada, your survival savvy, wit and knowledge, will carry with me forever. And finally, as always, to 'F'. You're with me through every project…my guiding light.

Thanks to all.

Jake

Editorial Method

◆

The purpose of this book is to provide basic survival information to you and those you love. It will present a basic understanding of survival should we be faced with an all-out nuclear, and/or biological/chemical attack. It is structured in a 'list' manner for easy to read, accessible information, that can serve as a guide. Further research is recommended for more in-depth scientific and medical views.

Introduction

◆

You are about to embark on a learning experience. *Surviving the Apocalypse*, is basically divided into four sections. Each section is designed to guide you through the process of setting up and planning your own course of survival. Every time I set out to write a new fiction novel which deals with apocalyptic scenarios, I create and devise survival plans that fit my characters and their personal situations and needs. What is inclusive in this book, is the knowledge and information I have collected as research material for those fictional novels.

You will find facts that are viable and true. Tactical survival guidance is based on research I have conducted, and information I have gathered through years of learning and talking to experts. Also, in order to aid you along, I have included a section which consists of survival sheets that may be useful for note taking, and course planning.

My manner of deliverance, at times, may be considered 'hard', but then so will be survival, should you be faced with doing so in an apocalyptic world.

PREPARATIONS

◆

Designed to get you headed in the right direction. Use this section in conjunction with the others to devise the perfect and best survival plan that you possibly can.

End of the World?

◆

It's a fall morning. You hate the thought of getting out of bed. The furnace hasn't been checked out, so you're reluctant to even put it on yet. In addition, the worrisome of an early high gas bill is in the back of your mind.

Let's face it, those of us who live in those wintry regions will wait until the last minute to kick the heat into high gear.

Your alarm clock goes off at the top of the hour. A newscaster on the radio is talking about some recent conflict. It's old news. You don't listen. You shut off the alarm, get out of bed and begin your morning routine for work.

No packed lunch, you'll grab a bite to eat from the local deli across the street from your office. In a rush, as usual, you kiss your wife good-bye before she trots off to work. The kids, they get on the bus, and you get in your car.

The forty minute rush hour drive to work is underway.

Fumbling with your coffee, you look for the new release from your favorite band and pop it into the car player. The third song is your favorite, and you search for it.

Finally, after sitting on the ramp for ten minutes, you get onto the freeway. Traffic moves slow, but it moves.

Hands tapping the steering wheel, your car vocals blaring out, and a long line of traffic before you; it's a typical work day.

Until…everything stops.

Your music winds down in a sour manner, the engines halt, the wheels finish spinning.

What happened?

You're so in your own world, you fail to notice everyone else is in the same predicament. Common sense fails to kick in, that perhaps something is wrong. You, with instincts, try to start your car. It makes no sound.

Flash.

White blinding light illuminates all around.

You shield your eyes. They burn.

Within seconds, they arrive. The howling, heat filled, winds of destruction. Five hundred miles an hour and carrying more than a hundred pounds of pressure per square inch.

Nuclear blast.

That is only one scenario. Does it mean the world has ended? No. An old school thought is that 'no one survives an all-out nuclear war'. Though there are no case studies to confirm or deny that theory, there is one thing that is for sure: The will of man.

It's pretty cut and dry. You live or you die. You falter to your fear, or gain strength and determination. You succumb to your ignorance, or you earn the knowledge needed to beat the odds and survive.

Which one are you?

I couldn't tell you how many times I have heard, 'I don't want to live after that'. My response is why? Are individuals that weak that they fail to have the will to start all over again? To face loss? Why wouldn't one want to continue on? If not to live, but to do things right. A second chance.

In my humble opinion, as cold as it sounds, modern man is weak. An attack would not even have to be a massive one to push man to the brink of chaos and panic. Society has evolved and handed mankind so much, that mankind no longer has the internal instinct to go and get it. At least not the masses.

I mainly deal with two different types of mass destruction in this book. Both are forms of terrorism and/or acts of war. Nuclear, biological, and chemical attacks. However, the survival solutions offered, and tips given, can be applied to other situations such as: meteor, natural plague, a technological breakdown of society, etc.

It is my goal to not instill fear, but to remove it. To inform. To help rationalize. To survive.

The Risk Factor

◆ ───────

In the beginning…there was nothing.

Then God did his thing, and after a step-by-step process that took time, you had a world. All right, it's a big comparison, but it's a good comparison none-the-less. Just like you wouldn't go and buy any car, or house without planning or looking into the purchase, you don't just jump right into your survival situation. You need to learn, plan and prepare. In this chapter of the book we discuss the groundwork.

There has to be groundwork with everything, a good solid foundation. This rule applies to survival. To do this, you have to remember the word, 'Risk' or rather, RISK. Now granted, you will think it the biggest oxymoron. Risk, the word signifying a dangerous situation, RISK, the acronym, being the possible means to a precautionary solution. But each in their own way, pave the road to being prepared.

Readiness
Information
Sensibility
Keen Awareness and Action

Now let's break them down…

Readiness

Getting ready is getting prepared. Readiness is planning ahead, making lists, thinking contingencies. It is also knowing what to do and when to do it. How to act plays a part. Knowing that in the drop of a hat, your survival plan can be successfully implemented.

Information

Just like the word says, be informed. You are already beginning that. Gain the knowledge that you need. Learn it. Research. Listen to the news. Get the facts. Learn the myths. Know the difference.

Sensibility

Keeping your wits about you before, during, and after is important. It is also important to be sensible while preparing. Think ahead, and think things through. Yes, while stocking your shelter with food, canned goods make sense. But think about it. Be sensible. Canned goods are heavy. What happens if you have to move your shelter?

Keen Awareness and Actions

Be extremely aware of your surroundings. Know when there is the slightest change. Be up-to-date on current events and possible war news. If there is a major crisis going on, don't choose to watch your favorite sitcom, watch the news. Have that roadmap of survival planned, and know the course of action you will take.

Planning Ahead

◆

Planning ahead is so vital. You do not have to be in the midst of a major military crisis to plan ahead. It can be a very peaceful time and you never know what will fall from the sky. Here are some suggestions in the steps for planning ahead.

Get a notebook. Taking notes is important, create a list of what you need. Break it down into appropriate categories, and then lists from there. Before you know it, you will have an entire notebook filled.

Assess the situation. Is there immediate danger? If so, implement an emergency survival plan. A viable, short term solution that will work for the time being. Once that is squared away, if nothing catastrophic has occurred, then resume a detailed plan of action. You have just created an emergency contingency.

Plan for survival and beyond. If you think a nuclear war is coming, don't just lay out a shelter plan that will last until it is safe to emerge. Plan further. Plan as if there will be no government handouts. Be sensible. Living off the land will not be an option of there are massive amounts of radioactive contamination.

Know what and where your shelter will be. If there is time to build one, and prepare for one, do so with care and precision. Then, plan a secondary means of shelter. This is a must.

Budget. Getting prepared does not mean, going broke. You can viably purchase the items that you need without spending a fortune. Discount stores are wonderful for things such as bandages, aspirins, paper products, cleaning supplies. Check out your list of items that you have compiled. Organize the list in order of priority, then lay out a shopping schedule. Devise a shopping scheme that works for your budget, but allows you to stock up in a timely manner. Do not be dumb. Buying one or two items a week will not do it. It is wise to do one initial 'stockpile' then fill-in from there.

Get others involved. You can do this by heightening their awareness, and recruiting their help. Teamwork is great. However, make sure only one person is an actual 'ring leader'. You may divide aspects of survival into expertise. For example, if your friend Sally is a nurse, you may want to put her in charge of medical supplies.

Visit the library, cruise the net. Learn as much about everything as you possibly can. From agriculture, to medical, to dentistry. Remember, you may have to be your own expert one day.

Follow the RISK factor. Ask yourself. Are you ready? If it happened tomorrow, would your plan be viable to go into motion semi-smoothly. Information. Have you learned basic survival skills? Are you up to date on the news? Sensibility. Are the choices you have made sensible? Can they realistically be carried through? Is your keen awareness heightened, could you be ready to 'act' on a moment's notice?

Remember, a pen and a notebook can be a very useful friend. Write everything down, like this book, divide it into sections. Never be afraid to ask questions, or 'read up' on subjects. Knowledge will be a saving grace, and ignorance your downfall.

Family Awareness

◆

No child, if they are old enough to understand, is too young to learn survival skills. For as much as we envision in the deepest sector of our survival fantasy that our family will all be together, let's face it, it is very conceivable that your child may be alone in the event of an all-out catastrophe. You want to make sure, for them and for you, that they are knowledgeable enough to know what to do. Though there can be reluctancy in bringing the young members of the family into the planning, it is advisable to do so. Like adults, and probably even more so, a well-informed child will not only fail to panic, but can be resourceful if they have the basic knowledge. As sadistic as it sounds, planning for the apocalypse, can actually be a fun family activity. Some things that you can do, to incorporate your family into planning are:

Inform your family, and make them aware of not only survival aspects, but of events that are occurring around them.

Always give situations and possible outcomes. For example, if threatening foreign country is doing something, or moving toward an invasion of a United States ally, inform your family of what could possibly come of the situation.

When in heightened military state, plan a nightly news time. Pay close attention yourself, and make up little weekly quizzes. Offer the family a prize to the one who gets the most answers right at the end.

Do not leave them in the dark on any aspects. Nuclear war, bio attacks, chemical attacks, may be ugly. But the more facts that your family knows, the more they can face the fear.

Have weekly survival meetings. Schedule them religiously, and plan the topics that will be discussed. Make it a new topic every week. During the meetings go over what you have, what needs to be done, and so forth.

Make a family project out of creating your own survival manual. And when it is complete, make a copy for each family member to have on them.

Any reference materials, books, survival guides, that you plan to teach your family from, have them handy at all times. Also, reading survival manuals out loud will promote understanding.

Have survival drills and tests. Pretend an attack has occurred. Test the family on reaction. Also, have a survival night, where you actually practice living, sleeping, and eating in the shelter. It will give younger family members a familiarity with the shelter surroundings, and lessen the chance of fright, if the time actually comes.

Devise a family contingency plan.

Make sure you go over the survival plan constantly. Have your family aware of every possible obstacle, and probable solution.

The Contingency

◆

You have read the word in this section. More than you realize, a contingency plan is vital. It is the means to a backup should your original plan go asunder. The main purpose of it is to have everyone aware in the event that everyone gets separated. Not only should you have a backup plan, you should also have a third backup plan. What happens if a family member cannot get to the shelter? Where would they go? What would they do? A contingency plan should not only include your immediate family, but those family members and friends that you would want to hook up with in the event of a catastrophe.

Devising an effective contingency plan would consist of:

> As mentioned prior, meetings. Meetings and discussions plant the seed. Constantly go over the contingency plan.

> Let's face it, chances of you being away from home at the time of an attack are great. For school age children inform them of where to go and what to do if they are at school. Plan a secondary shelter near your work. Malls. Any place you may be in the event of an attack. Know where you would go.

> Make a list of possible places you and your family would be in the event of attack. From that list know exactly where you would go and what you would do. Take a field trip, get your family familiar with the surroundings. Purchasing everyone in the family a pocket size compass to have on them is a good idea.

Allow them to know which direction to go, east, west, so on. If a compass is not practical, teach them to use the sun or the north star as a guidance for direction to the designated meeting place. For children, find what would be called 'safe places'. Since they would be too young to wander off or make it to a designated meeting place, they should have a safe place to wait until an adult or a member of your survival group can get them. If the shelter is sound, they may stay there.

Find a secondary meeting place, then a third. Assess your area. Is it at high risk? What is the likelihood that it would be totally annihilated? The secondary location and third, should be in low risk areas.

Write down the contingency plan if necessary.

Sample Contingency Plan

Below is a basic contingency plan devised by one of my fictional families. It can help you with your own, and is a contingency plan incorporated by the family to be implemented in the event of an all-out attack, with the world in disarray. The family is originated in Gaithersburg, MD. With five adult children, and many grandchildren.

The Slagel Contingency

A time limit has been set. Three weeks, in the event of biological attack, four in the event of nuclear.

The first meeting place is Gaithersburg Maryland, the family home. Home base. If you are in a bomb shelter, wait until it is safe to emerge. Then at the first available opportunity, make it to the designated meeting place.

Secure your children first. Make sure they are to stay put in their safe zones until you can get them.

Leave a note at your family home, or shelter. The note should be dated, and include your destination and health status. This will be informative for anyone searching you out.

If 'home base' is unaccessible due to destruction. More than likely, this would only be in the event of nuclear war or major bombardment. Home base should still be standing in any other circumstances. But if it is not, you are to go to meeting place two. Ripley, West Virginia. Should that be unaccessible, the third destination should be Oil City, Pennsylvania.

Once you reach a designated meeting place. Wait there. Do not leave. All family members have up to four weeks to arrive at the meeting place. If after four weeks, you do not arrive, we will assume something has happened to you. After four weeks, prior to initiating the long term survival plan, a small family search party will be sent out to look for members who did not arrive.

Leave notes! This is crucial. Before the remaining family members embark on the long term survival plan, a note will be left at the home base.

If you arrive late, and the family is gone. Do not take the note! There is still a chance other family members, or survivors, may seek us out.

       *   *   *

This, though from a fictional family, can be a map guide for creating your own. And the contingency, like everything else, is part of preparations.

SURVIVAL

◆

To beat the odds, to overcome obstacles, to skillfully secure your own existence. That is survival. The following section deals with the basic needs of survival. This section is geared toward helping you plan your immediate/emergency survival needs and beyond.

Shelter

◆

Shelter is one of the three most needed survival necessities needed in order to live. There are four reasons that shelter is vital. Shelter from the elements, from harmful substances, from dangerous attacks, and personal security. Shelter is also a widely debatable issue amongst experts, especially when it comes to a nuclear war. Many of the materials I have read bounce back and forth, and are far from consistent in what will work and what will not work.

First one has to assess the situation. What is the forefront danger, that shelter is needed? Realistically, if a plague has wiped out most of the world, or biological attack, and you are still alive, your home would suffice until you can implement your long term survival plan. However, you must be cautious. If you live inner-city, or in what once was a heavily populated area, your home or apartment may not be safe from looters, or those dangerously stressed out. In other words, elements are not your enemy, man is. Therefore, one can deduct, that in a biological apocalyptic scenario, shelter, or finding shelter, is not the major problem.

In the case of nuclear war, or war events where high explosives are used, shelter is essential. It is essential not only to protect from the blast, but from harmful fallout. In these events, pre-planning your shelter is vital.

But how does one determine what the danger is? What kind of catastrophe is coming? The best option is to plan for the worst. In my opinion, the worst is nuclear war. Nuclear war challenges all of your survival needs, from shelter to transportation, because of the high risk that most of everything will be destroyed.

So plan for the worst, you'll be well covered for any event.

Shelter.

The first thought that pretty much comes to everyone's mind is to run for the basement. However, unless your home is located about twenty miles from a major populated area, the average basement will be insufficient to not only protect from a full blast, but from radiation as well. If you are faced with an unexpected, and immediate attack of a nuclear blast, without any pre-preparations to your basement, it will give some protection. Once in the basement, here are some things you can do. These are based on a situation where the occupants of the home had no time to prepare for an incoming attack. It is my hope, that the next few helpful hints will be in vain, because you will have read the book and prepared. But, just in case you haven't. Here goes.

Quick Emergency Shelter in the Basement/Expedient Basement Shelter

First thing first. In an unprepared, un-stocked basement, block out the windows. This can be done with clothing, blankets, whatever you can use to block out windows and protect from explosions.

Shut off the gas line then shut off the incoming water line to the house. Doing so will keep the contaminated water from entering your pipes, and allow you to use the water already in your water heater for drinking.

Collect any and all plastic bags laying around. Garbage bags. Grocery store bags. They can be used for sanitation purposes.

Collect all clean laundry. Coats, tablecloths.

A cold cellar works best. If you have one that is ideal. Make sure you form a protective and thick blockade on the door. If you do not have a cold cellar…

Find the deepest corner of the basement. You will need to create a barricade wall to protectively corner yourself in. The earth will provide protection from radiation, but you need to create a barrier from anything that seeps in the house. A barricade wall can be made by using anything and everything you can find. Books are an excellent source of a blockade. An old shelf perhaps, if books are not accessible. Pack the shelves with linens. Another source that will work, would be to slide your washer and dryer over to make a wall. Stuff them with laundry and coats. Then create a roof with an old table or lawn furniture. Make sure you also place linens and such on the roof. This type of shelter will not give you standing room, but it will give protection.

In case of a nuclear explosion, following the blast, you have about twenty minutes before dangerously high levels of radiation appear through beta particles. Fill some sort of container with water. Take it into your barricade shelter. Radiation levels should be low enough in thirty-six hours to retrieve more water from the tank and to run upstairs to get food. Keep exposure out of your barricade down to under ten minutes. Remember, it is still not safe to leave the home, basement, or go outside. Radiation levels are discussed in full understanding details in, *End Times: Nuclear War*, section.

Basic Prepared Basement

The following recommendations will work well for a home basement if the home is located approximately twenty miles from populated area or, in the situation where you haven't had time to build an outdoor shelter, or proper indoor shelter. Having the following prepared and on hand will help in creating a safe makeshift shelter.

Don't throw out those old blankets, and clothes. Keep them sealed in plastic bags. These make wonderful window blockades. Also, enough bags of clothing can be a protective wall.

Old mattresses make good barrier walls. If you do not have an old mattress, visit a thrift store, they sell brand-new foam mattresses incredibly cheap.

If the basement is your shelter, then you must reinforce the main beam on the floor. Find the center pole to your home and run a secondary reinforcement beam. You still need to do this.

Again, if you have a cold cellar, this is your best bet. For rafter ceilings you need to reinforce them, then on the door to your cold cellar, attach a mattress for extra protection.

It is advisable to purchase or to make one to two bags of dirt or topsoil and keep them outdoors by each basement window. When a warning is given, you should incorporate the covering of windows into your emergency plan. In the event of unannounced attack, after the initial blast, as long as you stay within the safety frame, you can go out and use the bags of dirt to cover the windows.

Don't throw out old tables, they can be hurriedly placed together to form walls.

Making a Basement Shelter

If you are secure enough in the area that you live, you can successfully create an effective basement shelter. There are four types of shelters that I have learned of, and I would like to share. The cold-cellar-shelter. Again, I bring this up. But prepare your cold cellar ahead of time. Reinforce the ceiling. Pad, or put mattresses at the entrance door. Adding mattresses all around will also add warmth. The only downfall to a cold cellar shelter is lack of ventilation. But after a few days, it will be safe to open the door to allow a brief airflow in.

Divided basements are the best. Find the room in the basement most underground. Put pads or mattresses on the entrance door, and all exposed walls. If there is a window in the room, blockade that, but make sure it is blockaded with a removable padding. You may need this window for escape, and airflow.

The plywood shelter. This type of shelter is quick, easy, gives protection but little moving room. Find the deepest section or corner of your basement. Take a large plywood sheet and secure it at an angle against the wall, making almost a tent. Make sure the plywood has a slope to it. Against the sloped plywood, lean mattresses as a thick protection. Line them up across. Create a doorway with mattresses or sandbags.

The solid shelter bunker. This type of shelter is excellent, but takes time and money to create. Using the deepest corner of your basement, with cement blocks, you will create and build another wall, a new room so to speak, out of the blocks. You can leave a small section as a doorway, and use the mattress technique as a door. Do not rely on your basement ceiling as the ceiling to the shelter. You will need to make a roof for your bunker. This will and should be done with two sheets of plywood, and placed on top of that, mattresses for added protection and cushion. Very little standing room will be there, but safety is better than standing. In a day or two, it should be safe to leave your bunker to stretch in the basement for a few minutes.

Outdoor Shelters

Realistically speaking, most of you reading this are not going to go out and make what are called, extensive pool shelters. These are well constructed, underground, pre-designed shelters that take in account square footage, major ventilation systems and so forth. Though some of you may get inspired and for the heck of it, go build one. The downfall to building any outdoor shelter is that you must provide yourself with a ventilation system. Unlike with basement shelters, where the structure of your home affords some protection from harmful radiation, outside you are completely exposed. Especially if you place the shelter underground completely, common sense tells you ventilation is crucial. Ventilation is covered later in this section. Here are some examples of outdoor shelters:

What I like to call the wonder shelter. This is a well-designed shelter nearly completely underground. Are you a handy man? Have some time on your hands? Be the envy of your neighborhood and build this shelter. An eight by eight by five-foot area will be sufficient. You will dig an area as if digging a new foundation for a home. In essence you are building a foundation. You will make a brand-new basement in a hole. You can use, concrete, or five inches in diameter poles. If this shelter is completely underground, the roof should not only contain your entrance and exit hatch, but your ventilation hatches as well. It is advised to build the shelter up at least a foot above ground level. This allows for your ventilation hatch to be placed on the side, above-ground wall. Secure your exposed, above portion, and ceiling with sandbags. Your entrance door should have a sandbag as protection.

The trench shelter. Just as it sounds. This is protected primarily by sandbags. You will dig a trench, reinforcing the walls with sandbags, then create a roof, protected by sandbags, this can also be a partial above shelter as well. The trench will only need to be four feet deep, and above ground walls will be constructed out of sandbags.

The above ground sandbag bunker. This type of shelter is not recommended by experts. It does not have the blast protection being that it is completely above ground. It does, however, protect from fallout and radiation. This is a bunker or structure created with walls of sandbags, and a reinforced roof with dirt and sandbags. The walls of this aboveground bunker must be at least two sandbags, side by side, as width.

It is recommended for all-outdoor shelters, that you line the roof with plastic to prevent water from seeping in.

Ventilation

Ventilation is a vital part of any shelter. Basement emergency shelters, have a built-in source of ventilation. You open your barricade door, let the airflow in. But what about if this is not an option. Here are some very basic ventilation suggestions.

Your dryer vent hose. Keep this clean, fresh and longer than you need it. Pulling it into the shelter after a safety time frame will alleviate stale air.

The swinging hatch ventilation. This is for a ventilation system being placed on a side wall, It is not recommended for roof ventilation. Make a small hatch door. Two by two. Make sure it has removable thick protection and locks down when not in use. Like a swinging door, this hatch should be able to swing in and out. When refreshing air into the shelter, remove the barricade, unlock the vent, and allow for the door to swing in and out. This will create an airflow out and into the shelter when radiation levels are safe.

The vent hatch with removable fan. This is fairly easy. A small hatch that opens and allows for air exposure. Granted, there will be no power for a fan. Take apart an old fan, and rig it using a bicycle chain and peddle, or manual pull string so that you can manually make the fan go around. Turning one way will suck out the air, the other way will bring it in.

As with all ventilation systems, you need a secondary vent or outflow vent. Whether it be big or small, another small accessible opening should be in your shelter. When using a swinging hatch on a side wall, make the secondary opening on the opposite side wall.

If time permits, make a wood slate, a window blind style vent that blocks an opening in the shelter. Place a protective covering or shield over it when slates are not being opened to ventilate.

It is a good idea, no matter what type of ventilation system you create, to have a protective wall, or portion of a wall a foot or two before the vent. This will add some protection from radiation.

It is crucial, in all ventilation systems, that you must build a protective tent, either big or small, around the outside area of your hatch, or opening. This will prevent fallout, and rain from entering your ventilation system.

Stocking your shelter

About now, you're thinking, 'whoa, wait a second, none of these shelters are really that big, how am I gonna get all my supplies in there?' Quite simply…you don't.

With all that you need to stockpile as far as supplies go, there is no way you will get your supplies into your shelter. You will need to find a storage for the rest. This storage can be a corner of your basement. However, make sure supplies are covered with thick layers of protection. Another storage idea is to dig a small backyard trench and bury the boxes. A simple earth-covered hatch will keep them protected. Basically this is what you need in your shelter until it is safe to leave.

A two-week supply of food and water. This will take up most of your space, so think ahead when bringing food into the shelter. Canned goods, though not a great idea for long term survival, work well for short term shelter use. They stack well. Do not forget eating utensils, and a can opener.

One blanket per person. If it is cold, and feasible, try to minimize this by sharing blankets. Sleeping bags are bulky. If the shelter permits, the use of hammocks to sleep on work wonderfully. They can be folded up and placed out of the way during non-sleep times.

A battery operated clock. Designate a twelve-hour time keeper to keep track of time past. Do not put the batteries in the clock until you hit your shelter. Wristwatches will not work, and will stop due to EMP pulses following a nuclear attack.

A small amount of medical supplies. (See shelter med pack in the medical section)

A five gallon bucket with lid, for waste. Line the bucket with a garbage bag.

Soap, sanitizer, and disinfectant. Candles, or a light source.

Matches. Sterno, and Sterno stove for cooking. (See section on that)

These items will get you through the short term shelter period.

Shelter Warning

As with anything, there are situations that can arise in the shelter. Dangerous situations or things you need to look out for. They are:

Carbon monoxide poisoning. Use safe means for cooking, heating and lighting the shelter. Watch for signs of carbon monoxide poisoning.

Especially in the event of a nuclear attack, prepare for a lot of vomiting. This occurs not from radiation, but more so from psychosomatic illnesses. People generally assume they have radiation poisoning, and will display symptoms.

Shelter Fever. It is the same as cabin fever. This will occur. One way to deter it is to bring to the shelter activities. Puzzle books, books, a journal and pen for each person to write down daily thoughts will keep things in perspective. Be inventive, think of exercises, whether they are redundant or not, think of ways to move the muscles.

Illness from raw sewage. It is conceivable that the smell may get overwhelming in the shelter from sewage. Before risking the occupants of the shelter, realize that after a day, it is safe to remove the sewage from the shelter.

This concludes the shelter section of this book. I hope that you have found it helpful, and given you ideas.

Food

———————— ◆ ————————

You're probably wondering why the topic of food wasn't the first topic in survival. Pretty easy, food is a given. Everyone knows to get food and water, but shelter is assumed. Further in the Survival section of this book, I get into means to cook the food. This division deals with food you should consider stockpiling for your survival plan. Water is a division all on its own.

Food is essential. You do not have to be an Eienstein to figure that one out. But believe it or not, people are pretty dumb when it comes to stocking up food. During the Y2K scare, I had the privilege to watch what people purchased. It was ridiculous. Where were their minds at? They bought foods that didn't need heated, they needed cooked fully. And the cooking time for these items would exhaust a lot of their cooking resources. I had to wonder if any of these people even thought or planned on a cooking resource. I vividly recall one woman in line with seventy-five cans of tuna. Seventy-five? Granted, tuna is a source of protein, it does not need cooked, but seventy-five cans. That tuna would be useless, because the occupants of the shelter would tire of eating it, therefore would only 'pick' at their rations, not consuming enough to sustain them.

You do not need to eat a lot of food to live, you just need to eat the right foods. And remember, just because you are in a survival situation does not mean you cannot enjoy what you eat as well. I hope that the following items listed help you in determining what you should stock for your survival needs.

Recommended food items for short term shelter

As stated earlier in the book, canned goods are not a good idea for long term use. They are too heavy for when you move your supplies. However, if your home shelter is the place that you intend to make your long term living place, then canned goods will work. Before you plan on using your home as your survival base for the future, determine the growth possibilities that you have. Also, canned goods are great for the short term shelter stay.

Choose your canned goods wisely. Corn is a filler and offers no nutritional value. Though some cans of pasta are an excellent source of nutrition, they can be pricy. Pick vegetables, beans, and fruits. Meats, stews in a can are great for shelter.

Though cans are not recommended for long-term, there are canned goods that can be stockpiled. Potted meats, Vienna Sausage, are in very lightweight cans and they offer protein.

Other protein items are peanut butter and dry beans.

Oatmeal. Rolled oats. The instant kind is awesome. Use for protein and a filler.

Dry cereal. Make sure you double seal it in a plastic bag.

Canned cheese. This is a fun food. Crackers will last the approximate two weeks in a shelter as well. This can be a great diversion or snack while waiting for the main meal.

Baking soda. I had a hard time determining what category this falls in. So I placed it here. It works for teeth, antacid, to put out small fires, and smells.

Instant beverages. Coffee, tea, juice, Tang, evaporated Milk. Condensed milk. You may want to get the sweetened canned milk for the kids.

Some candy, hardtack and chocolate for treats.

Here's the one I like. We as modern man have a huge and wonderful advantage over our ancestors when it comes to preplanning our survival and stocking. We have…Ramen

Noodles. Yes. Don't laugh. In the 1980's, Ramen noodles were a delicacy, and were priced very high. Now they are virtually free. Not only are they lightweight, they cook fast, are tasty, use little cooking time and water. They may have little nutritional value, but all that can be changed by adding to the noodles. Sausage, vegetables. For easy carrying and storage, you may take the Ramen from their package, break it, and place multitudes of them in one very sealed plastic bag.

Long term food stockpile

Along with the above suggested, the following items should be purchased and on hand. These are the items you want to store in a safe place so you can access them after your emergence from the shelter.

Flour. Flour is a long term item. You will not need to bring this into the shelter with you. Let's face it, you won't be a Better Crocker in the two-week wait until radiation levels drop. Make sure flour is sealed in plastic bags.

Sugar, salt, seasonings. Sugar and salt have their obvious reasons, so why seasoning? You may need the seasoning for future dehydrating. Also, tons of salt. If you end up hunting. Salt preserves food.

Rice. Beans. Pasta. Ramen Noodles.

The prepared dehydrated items. Make sure these are also sealed in plastic.

Seeds. These will be for future growth. You will need to start seedlings. Which are pre-potted plants. The reason being is any soil will not be a viable source to grow food. You will have to remove at least six inches of dirt. This can easily be preplanned. Start saving large jars. Mayonnaise, pickle. Purchase one large bag of top soil. These can be the starts for your new farm. They take minimal light and can be started indoors during cold months.

Keep in mind when thinking long-term. Stock up! It may take a while for you to get the agriculture hang of things. Realistically, a productive growth system may take years to kick into gear. Especially if you have no farmers.

Dehydrated Food

Dehydrating foods ahead of time is your best bet. There is very little you cannot dehydrate. If you dehydrate one or two items a week, you are already stockpiling food for a long time. For example, even dehydrating three pounds of beef, and making tomato rolls, gives you about eight days' worth of food. Dehydrated food can be eaten alone, or mixed. I recommend mixing the items. Dehydrated food can last indefinitely when stored in the freezer. It will last nine months or longer (depending on the food) if kept sealed and in a dry, cool, place.

A dehydrating machine is a wonderful tool, however many of us do not have one. Dehydrating is so easy. It is simply done by setting your oven on a 140-degree setting, propping your oven door open slightly with a wooden spoon for airflow, and allowing the items to stay in the oven for anywhere from ten to twelve hours. You know food is fully dehydrated when it feels leathery to the touch and no moisture is seen or felt.

My best advice to you is to experiment with dehydrating. Try things out. All of these items can be reconstituted with very little water. You do not even need to boil the water. However, it makes things move faster. Below are some suggestions on what you can dehydrate. Most of these can be done in a dehydrating machine, but these recipes are designed for individuals without a dehydrator. All items should be placed in airtight plastic bags. Divided in small helpings.

> Tasty Beef Jerky
> Three pounds of round steak.
> Pound out round steak until paper thin.
> Marinate strips in Soy sauce, garlic, and sugar.

Lay piece flat out on cookie sheet
Dehydrate for twelve hours at 140 degrees. Divide into three
   packs. (One pound packs)

Ground Beef Bits
Brown three pounds of ground meat.
Season to taste.
Drain all fat.
On ungreased cookie sheet create a layer
Dehydrate for ten hours
There will look like there is nothing left. But when reconstituted,
   they expand.

Chopped Onions, Carrots, Peppers, Celery
Do these separately because these are a sight and judge item.
   Meaning you will have to start to check them after eight hours.
Chop vegetables thin.
Spread evenly on ungreased cookie sheet
Dehydrate for eight to ten hours.
These items can be tossed into cooking water with Ramen noodles!

Soups
These are fun, popular, and extremely tasty!
Two cans of a cream soup. Cream of chicken, celery, tomato,
   potato, any will work.
Grease a cookie sheet lightly with cooking spray.
Spread the thick soup in a thin layer across the pan
Dehydrate for ten hours (sometimes less)
When done it will appear like a colorful sheet of rubber
Cut into one-inch thick strips. Strips should be the short width
   of the cookie sheet
Roll up the strips individually

Tomato Leather
Three cans of tomato paste.
In a bowl mix garlic and season into paste
Spray cookie sheet with cooking spray
Spread tomato paste thin on sheet
Dehydrate for ten hours.
Cut into inch thick strips the short width length of cookie sheet
Roll up strips individually.

The Tomato Leather and Soup Strips can be reconstituted in six ounces of boiling water. These also can be eaten as is. Also, a little roll added to Ramen helps change everything!

Short Term Shelter Food Amounts
The following is just a sample list of what would be needed to survive a two-week stay in a barricade short term shelter during a nuclear holocaust. This is based on a family of four. Remember, you will be rationing!

Twelve gallons of water
Two jars of peanut butter
Four boxes of crackers
Two cans of spray cheese
Two boxes of dry cereal
Four cans of beans
Five cans of pasta
Four cans of carrots
Four cans of another vegetable
One canister rolled oats
Ten packs of Ramen
Four cans of tuna
Five cans of Vienna sausage
Jars of instant coffee, Tang
Fours cans condensed Milk
Four cans of fruit

This basic menu for twelve days will provide rationed water and three light meals enough to sustain. One cup of dry cereal or oats for breakfast. Ramen noodles with four peanut butter crackers for lunch. An example of dinners may be: Pasta, canned fruit, and four cheese crackers. Vienna Sausage, beans, crackers.

Emergency Survival Food Kit

The following are items that will go into an emergency food kit. You will find out at the end of the survival section what this kit is used for.

Kit includes:

>One, one pound pack of beef jerky
>Two cups of dry cereal placed in two plastic bags
>Two packs of Ramen noodles combined, and broken in one bag
>One cup of oats
>Seven strips of Tomato Leather
>Seven Strips of soup
>One sealed fresh pack of crackers
>One very small jar of peanut butter

This pack will weigh no more than two pounds, and will fit in a large shoe box. Always remember, when sealing items in a zipper plastic bag, squeeze out all air prior to finalizing the seal. Plus, cover the dehydrated packs of food with tin foil.

Water

———————◆———————

Water is in a section of its own because a lot is to be said for water. Aside from using water to drink, water will be used for cooking, sterilizing, cleaning, and agriculture. However, this section is going to deal with storing water for immediate and beyond. Long term water use for agriculture and so forth, should be a long term plan. Such as finding a fresh water lake, or well water. Although most individuals may consider a 'water' section not necessary, a section to inform is crucial. Keep in mind, in a nuclear war, more people will die from waterborne illnesses than killed from a blast and radiation combined. That is a scary fact.

One of the most debatable issues is how much water does one need to survive? Experts say the human body, on average, needs to consume five quarts a day. Or twenty glasses of water. I know I do not drink that much water a day. To survive, the human body needs to consume enough liquid so as to urinate one pint a day, or sixteen ounces. One pint ensures proper filtration of body waste. In every source I have consulted, all of them agree, that three pints per day would suffice. Especially when dealing with low levels of food. When drinking low amounts of water, make sure salt is added, even just a little, this will help the body sweat any impurities out.

Civil Defense recommends that you should store fifteen gallons of water per occupant in the shelter for a two-week period. A gallon of water per day, per person? Not only is that ridiculous, and more than what the human needs, but it is a waste of valuable shelter space.

You don't need a lot of water to survive. History, and circumstances prove people have lived for weeks on less than a glass of water a day. Now I am not telling you that you should make the occupants of your shelter suffer from painful thirst. I am telling you that not that much is needed. Judge for yourself. I am sure, with initiative and research, you, the reader, can design a sophisticated water system, but most of us are not that industrious, nor have the resources, so I offer you my easy advice.

If you have no time to prepare a shelter, and you thought fast and shut of the incoming water valve to your home. Right there, sitting in your basement are twenty gallons of drinkable water.

If you have time to prepare a shelter, twelve gallons of water placed into your shelter area (Cornered off section of basement) will suffice. Keep extra bottles and gallons of water near by your expedient shelter, or planned shelter. Keep it protected and covered, but close enough to retrieve if need before the two weeks are up. I keep saying two weeks. With Nuclear war, based on an all-out, 6000 megaton impact on the US, radiation levels should fall to safe levels within two weeks. Less time if you live in low risk areas. This means you'll be able to leave your shelter for extended periods. However, you will be able to leave earlier, and safely, for shorter periods.

Do not go out and buy bottles of water. Don't be ridiculous. Start buying milk by the gallon. Soda-pop bottles. All of these can be washed out and filled, then stored in the basement, or near the shelter. Any water not brought into the short term shelter should be protectively covered. Fill these bottles up constantly. Empty and refill them every eight weeks so the water is the freshest it can be at the time of attack.

Make a well. This is done by building a pseudo, outdoor swimming pool. A trench lined heavily with plastic will work. Fill this, then cover it with a layer of plastic, then an old door. Cover the door with dirt. In cases where you dig a trench or make a well, please recall where you do this.

Water cans. Garbage cans lined with plastic, then filled up make wonderful water storage bins. Especially for use after you emerge from the shelter. Make sure the water is covered and the cans insulated for protection.

Water Warning

Clear, no smell, don't be fooled. The water will and can kill you, especially in the event of a nuclear war. Here are some advisory tidbits for using water in the immediate and beyond.

Don't be stupid. Rain, snow, water draining from a roof, are contaminated water sources.

Boiling water may remove bacteria, but it will not remove radioactivity.

Water from deep wells and reservoir tanks are your best source for water supplies following an attack. Especially if these are covered and no fallout has gotten in them.

Deep lakes are a source of fresh water. You must go toward the middle of the lake to retrieve fresher water.

Spring water, though naturally filtered can still be contaminated. Assume it is, it doesn't hurt to follow precautions that I have listed later in this division.

Creek water is a no-no. Especially if muddy.

You may cringe, and wonder how you'll get water when your immediate supply runs out…don't. Don't panic, and don't cringe. There are ways to remove ninety-nine percent of radioactive contamination from water. The safer the source, the more effective removing radioactivity is. However, you must think long-term and beyond. Be resourceful, the suggestions I make are credible, but time consuming. Learn now, research now on more sophisticated ways in the long term future to filter war.

Making Water Safe

Although not a hundred percent, there are ways to help make your water safe. Here are some ways that I have learned.

Boiling water for ten to twenty minutes will remove a lot of common bacteria. But not radiation.

To disinfect water from waterborne illnesses, bacteria, and/or biological attack, there are three effective ways other than boiling. (Which I recommend boiling as a back up anyhow). The first is simple household bleach. Yes, that's right. However, check the label! You can only use a household bleach that states the **only active** ingredient is, sodium hypochlorite. This must be the only active ingredient in the bleach and the percent of sodium hypochlorite should be 5.25%. Use one teaspoon per ten gallons, or two medicine droppers per quart. If water is muddy, double these amounts. Stir, then wait thirty minutes before drinking.

The second method is 2% tincture of iodine can be used for disinfectant. Five drops per each quart, ten if the water is muddy. Stir, let stand thirty minutes before drinking.

The third method is a standard commercial water purification tablet. You may purchase these and follow the directions.

Removing Radioactivity

Yes, it is possible. There are two ways to do so, and do so effectively. As mentioned before, only ninety-nine percent of radioactivity is removed, it is still better than having the water full radioactive, right? Before I proceed, here is a term you will need to know: Clayey soil.

Remember when you were a kid, and you used to dig in the backyard. Remember how once you got about four inches into the ground, the color and texture of the earth changed. Usually, it went to an orange, or orange brown, and it was almost a clay substance. That is clayey soil. This is found about four to five inches from the earth's surface, and is a safe earth from fallout and radiation. Now that you know what that is, we can go on. Here are the two methods I know that can be used in utilizing nature:

Water Filter System
You will need a five-gallon tin can, bucket, or one of those large restaurant size cans. Must be metal. Two large wash-cloths. If wash-cloths are not available, a thick shirt cut will work. You will also need a 'catch' container to catch your water as it drips from the filter system.

First, poke five screwdriver size holes in the bottom of the can. Make sure the metal protrudes outward. Then layer the bottom of the can with washed pebbles, or stones. Next, lay the wash-cloth over the stones. On top of the first washcloth, place about six inches worth of loosely packed clayey soil. Overtop of the clayey soil lay the second washcloth. Secure the wash cloth around the inside perimeter of the can with a couple small stones. Place the can over the 'catch' container. Perching the can on two sticks or some leveling device will work, then start to pour your water through the top of the can. The water may pout sluggish. If it backs up, remove the top washcloth, rinse it, and that should do it.

Settling System
This works really well too. Fill a large five gallon bucket with the water you want to remove the radiative material from. Next, take three double fistfuls of clayey soil and crumble it into the water. Stir with a stick until you see the clayey soil spinning about and floating about the water. Allow the water to sit for six hours. The clayey soil will settle and bring with it all the radioactive material.

Better Safe than Sorry

I know it is time consuming, but I say it never hurts to doubly back-up yourself. In this case, quadruple yourself. I recommend doing more than one of these for any questionable water supply. I recommend settling the water, then filtering it. After filtering, add the bleach or iodine, then boil. Again, time consuming, but you will be covered and would be more secure in drinking the water if you feel you have done all you could do.

Medical

◆

Brief but important, the medical subject is one needed to approach. I'm sure, if you are a doctor, there are things you can get readily. Antibiotics, anti-anxiety tablets, sleeping pills, potassium of Iodide (For radiation sickness), and other things. But getting ahold of those items, for the average person is tough. So, I have placed together a list of items that you, as a layman, without the means of having the power of a prescription pad, can put together and stockpiled. All of which will be very beneficial. Keep in mind, all over the counter medication will eventually expire.

If you can find a means to getting Penicillin, do so. There are various places on the net and through Militia sites that are reputable and can get you squared away with these. A few hundred tablets will work.

Aspirin is vital. Very vital. Not only is it very good for pain, but one aspirin, 80 milligrams, placed on the tongue during the onset of first symptoms of a heart attack, can decrease the effects and severity of the heart attack. Plus, aspirin has a shelf life that can be used post the expiration date.

Acetaminophen. This should be your main source for pain and fever reduction.

Ibuprofen. Try to use this for swelling. This will come in handy for sprains, and breaks.

Premenstrual Syndrom, or Menstruation Relief medications for obvious reasons.

First aid supplies including: Bandages, antibiotic ointments, bandages, ace bandages. Splints of all sizes and shapes can be purchased at the pharmacy. Slings. Secure all bandages in a dark plastic bag, and seal it. If this bag does not make it into the shelter, make sure it is protected from fallout.

Cold medications. Including, cough suppressants, decongestants, expectorants. Many respiratory infections will spread fast and furiously in the close proximity of a shelter.

Asthma relief medication.

Anti-diarrhea, anti-nauseousness, antacid medications.

Sleeping pills. These will assist with anxiety and sleeplessness.

Cortisone, hemorrhoid medication, suppositories.

Iodine, peroxide, alcohol. Though it is proven that peroxide holds no medicinal value, remember the bubble action works to clean out wounds.

Bourbon. What? Yes, that's right, whiskey or bourbon. Not only is this a means of disinfectant for wounds, but an excellent calming agent. But be cautious. Alcohol dehydrates you.

Vitamins, especially ones with potassium.

A complete medical dictionary. Not only will this be a useful tool in assessing situations, medical dictionaries have nursing appendixes in the back that tell you what to do as far as treatments go. Plus they have a wonderful fist aid section.

The best, and foremost important thing you can invest in is a first aid homeopathy kit. There are many homeopathy pharmacies that you can mail order from, you may even have one locally. I am not talking about the kind of homeopathy that you get in the local drugstore, I am talking authentic homeopathy kits. You can get a forty vial or eighty vial kit. Each vial or tube contains hundreds of tiny pellets that can be placed under the tongue. A single pellet is a single dose. Or eight pellets mixed with one ounce of distilled water with a dosage of a dropper full.

The homeopathy vials contain natural treatments for everything from the flu, to excessive bleeding, to hangovers. The bigger the kit, the wider range of treatments. Each kit comes with a sheet that tells what each vile can cure. Though pricy, I highly recommend getting a kit. They are all natural and do not expire. Plus buying a homeopathy medical book as a companion works. I speak from experience when I tell you the miracles and wonders of homeopathy.

Short Term Shelter Supplies

It is advisable to keep as much of your medical supplies in your shelter as you can. The rest should be very well protected and shaded in the event of a nuclear attack. The homeopathy kit is six inches by six inches and takes up very little room.

Emergency Medical Pack

Yes, this is one of those things whose usage will be explained at the end of the chapter. A small bag or pouch about six inches wide should be used as a medical emergency pack. It should include:

> Placed in a small sealed plastic bag: Eight aspirin, eight acetaminophen, eight ibuprofen, four sleeping pills, four anti-diarrhea pills. (Please separate so you know what is what) if you can purchase trial sizes a head of time, all the better.
> A small roll of gauze and tape. A small handful of band-aids
> A small tube of antibiotic cream.
> A small one ounce bottle of bourbon. Airplane size.

Sanitation

◆

A division in itself because sanitation is so very vital Aside from the obvious personal items, one must plan ahead. In close proximity quarters such as a shelter, sickness and illness can spread with the tiniest of germs. Every effort should be taken to maintain the best of cleanliness from personal hygiene to shelter hygiene.

The Shopping List

I would assume, most people would be able to judge what to get for the shelter and long-term survival without any guidance from me. However I have provided a list of things you may or may have not thought about:

> Personal hygiene items. Bar soap, toothpaste, shampoo, toothbrushes, disinfectant mouthwash, hand cream
>
> Household cleaning products. Chlorine bleach. Dish soap. Laundry soap. Pine disinfectant. Air fresheners
>
> Feminine protection, diapers (If needed), toilet paper
>
> The amazing liquid sanitizer. It is that clear gel that you place on your hands, rub your hands together and it disappears as it dries. This is wonderful and a must.
>
> Rubber gloves for waste handling. And speaking of which…

Waste

> As much as we want to avoid this topic it must be broached. Human waste is a given and will occur. Unfortunately, in extremely stressful situations, coupled with possible radiation

poisoning, human waste will reach its extreme. Here are some suggestions on preparing for conquering this delicate subject.

If you are in an emergency situation with no time to prepare. For male urination a laundry bottle will work, plus carry a certain amount of odor protection. This can be capped, then drained out when full into a basement drain. Some things you may have in your basement that can be a makeshift toilet are buckets, paint cans, large cooking pots, bowls. Place expedient toilet in the corner of shelter and try to section off with a blanket or sheet.

The ultimate makeshift potty. If there are no means to buying a porta-pot, or commercialized mobile potty, there is a way to make one if you are preplanning and pre making. Take a five gallon bucket and line it with small, sturdy kitchen garbage bangs, place a small amount of water in the bag. Place one commercial sanitizer tablet in the water. A lid will work, but if you have time, you may want to purchase a toilet seat to place on top. Partition off a section of the shelter for privacy. Depending on frequency of use, empty every other day.

You may want to use a similar method with a smaller bucket and lid for a vomit pot.

Helpful Sanitation Shelter Hints

Prepare a schedule for plastic bag waste removal. Make sure the removal person wears gloves. Rotate who removes the waste bags. Wash hands or use the sanitizer religiously. Keep in mind, a lot of germs can be destroyed through the vigorous rubbing of your hands together with the smallest amount of soap.

Prepare for smells. In the small space smells will not be pleasant. A spray disinfectant may help.

Do a daily clean up chart. Make it like clock work one person sprays and wipes down all items in the shelter with cleaning solution. This will help with smells and germs.

Wipe down the seating area of the toilet frequently and use paper for when sitting, or do not sit directly on the pot at all. Also, encourage males, and younger occupants not to 'miss'.

Cooking and Heating

◆

The last thing you want is for everyone in your shelter to die from smoke inhalation or carbon monoxide poisoning. So once, again, I preach ventilation. Make sure there is some sort of ventilation in your confined area. It is my hope that right now common sense has kicked in and you realize that you cannot build a fire in a small shelter area. So use of flames and fuel should be limited. So, then you ask, how do you eat, see, and stay warm. It's not so difficult. Especially if you preplan.

In an emergency, unprepared shelter situation most foods can be eaten without cooking. Canned goods, dry cereals, if you grocery shop on a regular basis you should have the items, or at least some, in your cupboards.

In wintry months, you will have to rely on body heat, blankets, and layers of clothing. Layer clothing on the body, thinnest to thickest. Also, start saving newspapers. Newspapers are an excellent source of insulation. In a small shelter, you will be amazed how much heat a candle or two gives off, along with small cooking devices.

Watch out for commercialized gas grills, gas lamps, kerosene lamps and heaters. In closed in spaces, they produce gases that will cause headaches and even death.

Use candles and lights sparingly. Bargain stores have battery operated lights that last a nice amount of time, and can be used intermittently with candles. But it is advised, during winter months to use candles.

Always cook, and use candles and lamps near shelter doors to promote ventilation.

Be smart. Be safe.

Makeshift Lights and Stoves

Jackie Stove

My invention, basic and pretty simple. It was tested by a friend, and tested by myself. It works wonderful. Sterno can be purchased at any store. It is my recommendation that you start saving Sterno. They burn slow, produce a lot of heat, and can cook eight ounces of water in less than a few minutes. The Sterno stove is constructed with a small boiler grate. Two bricks and a Sterno. Between the two bricks place the Sterno. Rest the grate on top of the two bricks. There you have it. Now the trick is to not use the Sterno for very long. Make sure you use a pot that has a tightly secured lid. After water or contents comes to a boil, extinguish Sterno, cover contents. Pasta will cook in fifteen minutes in a sealed pot.

Insulated Flame-less Cooker.

This is a great tool for items that need a lot of cooking time, plus it supplies a harmless means of heat. You will need a large bucket or basket, non-plastic, preferably wood. You will also need a handle-less pot with a secure, tight lid. Pack the large bucket with newspaper, but leave room for the handle-less pot. After the contents of the pot have boiled for one minute. Cover pot, and place it center the bucket of newspaper. Pack more newspaper tightly around it. The newspaper will keep the heat in and keep the contents cooking. It is like a slow cooker and will take time, but it works and provides heat.

Light Lamps

For heat and light. These emit little gas and smoke. Partially prepared ahead of time, these lamps burn nearly endlessly. Eight
hours will consume one ounce of oil. You need one jar, two
nails, or something for an anchor, a six to eight-inch wick, very
thin line wire (Such as used for bread ties) and cooking oil. Tie
the two nails together with the wick, leaving a six inch length of
wick. Wind the wire up and around the wick for sturdiness.
Making the wick erect. Place the nails as an anchor at the bottom of the jar. Holding the wick, carefully fill jar with oil leaving
three quarters of the wick out of the oil. Also, drilling a hole in a
small block of wood, and running a wick through that will create a floating wick for your lamp. It is advised to make the wicks
and anchors a head of time. This can be a fun family activity.
Something to do one night during a survival meeting.

Tools

◆

Surprisingly the concept of tools fails to cross a lot of people's minds when it comes to preplanning their survival. One would think that things such as a knife, would be top on a list. However, during the cold war when a lot of surviving research was completed, plus on a list of surveys I handed out to people, the number one forgotten item was…a can opener. As hard as that is to believe, it is true. Sometimes the simplest thing, escapes us.

I have compiled a list of tools. Tough the list is not long, all of these have multi uses. Most of which can be stored in a safe place, and not necessarily needed in the shelter. So grab an old trunk, or plastic bin and start collecting.

Tools

Dad or Grandpa's toolbox. You know the one you make fun of because everything is so ancient and old, and out dated. Grab it, save it, that will be your most useful asset. Old tools, outdated by modern electric equipment are sturdy and take you back to basics. This toolbox should have the standard wrenches, screwdrivers, hammer.

Other metal items such as: Screws and nails. Saw, crowbar, hunting knife, and shovel. Also a knife file.

Pocket knife, bottle opener, can opener.

Rope and twine.

Scissors, and a sewing kit.
Eating utensils.

Primarily the items mentioned here are tools. Basic tools. There are a lot of things you may consider tools that are covered in the 'Miscellaneous' section. Also, as you probably noticed by now, I don't cover weapons at all. Common sense, and other resources can teach you about self defense, that is not the intent of this book. For hunting, guns are not necessarily needed. Sharp stones, filed down can make great arrows. Plus, if martial law has kicked in, you may not want to take a chance of getting taken into custody because you are bearing arms. So use weapons and carry them not just with personal caution, but with lawful caution as well.

Miscellaneous Items

◆

Ah, the illustrious 'Miscellaneous Items' section. Most people may even read over this, figuring Miscellaneous Items is another term for 'useless information'. Untrue. What is covered in the division of the chapter are items that you may or may not think about. You may see items previously mentioned in the 'Survival' section. No, that is not an error, they are just items that need reiteration, and items not readily thought of. This is a good list to review, because this is one of those sections in the book, while reading, you'll find yourself saying, 'hmm, I never would have thought of that.'

The Big Miscellaneous List

Garbage bags, sandwich bags, plastic bags. These items are crucial and have multiple uses. Aside from the obvious, bags can be placed over feet and hands for warmth.

Aluminum foil. This is wonder wrap. You would be surprised what you can use this for. Insulation. Reflection for light. A makeshift antenna for a radio.

A wind-up clock to keep track of time, and a calender.

Notebooks, puzzle books, reading and writing materials. This will help pass time in the shelter when all are bored.

A short wave radio if applicable. If not, even a transistor radio will work. With the airwaves not flooded, it may pick up signals.

If there is a musician in the shelter, encouraging this person to bring the instrument (obviously this doesn't apply to drummers and piano players). Music is very soothing, and it may be needed.

A bag of sand for the shelter. Small to be sprinkled over any damp areas or spots where body fluid may have been projected.

Batteries of all sizes.

A few rolls of duct tape. Most men call it the universal 'fix it' tool anyhow.

Items mentioned previously: Toilet paper. Sterno. Can opener. Scissors. Candles. Newspapers

Matches. You should truly buy a stock of them.

Mosquito netting to cover shelter door if bugs become a problem.

It may be a good idea to have a tarp or a large rubber sheet. As cruel as it sounds, there is a chance that someone may die in the shelter. Also, if they are suffering from sickness, they may be expelling multitudes of contagious and offensive fluids. Placing them on a rubber sheet will aid in keeping the shelter free from excretions, and aid in cleaning up should this person pass on. Remember to roll the body in the covering, and remove the body from the shelter immediately. Burial can be worried about later. You're alive.

Personal hygiene items such as soap, toothpaste, hair brush, and for women hair bands. In a low water situation, dry shampoo that brushes out will help. Because it is an apocalyptic situation doesn't signify bodily cleanliness goes out the window. Especially if the means are available. Remember, just because the world went to pot, doesn't mean that you should too. Keeping clean is not only recommended for sanitation purposes, but it will make you feel much better. So go on, brush your teeth, wash your face and comb your hair. Defy what all those post-apocalyptic movies depict what survivors look like.

If you can get a hold of a sump pump, or small water pump. You will need to makeshift a longer hose. This will be extremely useful for pulling gasoline out of underground tanks. Because without electricity, you cannot pump gas.

And last, but not least important: Don't forget to bring your survival plan and books.

Emergency Survival Plan

———————— ◆ ————————

It is extremely important, and I cannot emphasize enough, the value in preplanning. You very well may be stocked up, the contingency in motion, every item on the list, checked and re-checked. You know what to do…you think. What happens when the time comes? As you read further into the book, you will see, that most of the shelter, and emergency precautions are geared toward a nuclear holocaust. Mainly, that is because biological and chemical warfare has no warning. And even if so, you can't see a virus in the air. As far as chemicals go, if you smell them, you're history anyhow, or pretty down and out at the least. Reiterating why I focus quite a bit on Nuclear War. That is where your gravest survival needs will fall. Below you will read a pseudo attack plan. It is created regarding a family of four: Karen, Chuck, Robbie, and Tammy Clinger. They are the preplanned, stocked and ready family. This is their master plan that they have done drills on, and practiced. This is plan devised, should they all be home during an attack.

Clinger Family Emergency Routine
>Listen for the warning sirens, or announcement on television through the emergency broadcast system.
>Instructions are given to the children, Tammy and Robbie, that if they are in a two block radius of the house, they are to take the backyard route home. If they are further, do not chance it, run to the nearest shelter or house.

At that moment, with a safety time frame of fifteen minutes, in the event that one or both of the children are outside playing, Chuck will gather the children, and Karen will take over his duties until Chuck arrives back.

Karen shuts off gas and water.

Chuck will go outside, and cover basement windows with sandbags that are outside of the basement windows.

Karen will gather all accessible, and readily available blankets and take them to the basement.

Robbie will block out basement windows from inside.

Tammy will do a quick check to make sure shelter supplies are in basement shelter room.

Robbie and Tammy will double-check the covering of extra shelter supplies left out of secure room.

Karen will place any fresh fruit items from the kitchen into plastic bags. These will be the first items eaten in the shelter. She then will go permanently to the basement, and gather children into sectioned off, pre reinforced basement shelter.

After covering basement windows, Chuck will proceed in the home. He will get the huge bag of dog food, pour some in a dish, then leave the bag on the floor opened. Chuck will then fill a large pan with water.

Chuck lets the family pet, Spot, into the home. Spot cannot go into the shelter, unfortunately. He is a mouth to feed and will use up valuable fresh air and oxygen.

Leaving Spot on the first floor of the home with a prayer, Chuck heads to the basement. Secures all FIVE bolts on the basement door, then proceeds with family into shelter.

All if this should take no more than eleven minutes.

Emergency Pack

In the previous medical and food sections you were given instructions on making emergency food packs and medical packs. I had said earlier that you will find out what these are needed for, here is the paragraph or two you have been waiting for.

Suppose you are driving. Suppose you are far from anywhere you may have designated to go, or even close to that basement you charted out in your contingency plan. What then? Obviously, if you pick Donner Funeral home as a viable 'If I am at the mall' contingency shelter, you cannot pre-plan and pre-stock Donner Funeral home. So basically, you have shelter. What about food? Medical? Anything else. Hence, the survival packs.

I preach time and time again to preplan and prepare. This is a must! In the trunk of your car, you should have one emergency food pack, one medical pack, a blanket and gallon of water, per person that would be in that car. Plus, one roll of duct tape. Each food pack will give a single person enough rations for two weeks. With rationing, they can make the water last.

It also may be a good idea to make mini food packs for the office or school.

END TIMES

◆

I believe this famous quote says it all.

> *"I do not know how World War III will be fought. But World War IV will be fought with sticks and stones."*

Albert Eienstein.

Chemical Weapons

———◆———

I am placing my small introduction to 'End Times' under this first division. It was extremely difficult for me to determine which subject I would put where. It had crossed my mind that my reader would choose which topic they preferred to read first, but then there would be the readers who would read it in order. I knew right off which topic I would place last. However, that still left three. Chemical, Biological, and Nuclear War.

I asked my son which one he felt I should place first. He shrugged and said he didn't know which one should go first, but he knew which one he would put last. Well, since his final choice differed from mine—his reasons being that his choice was much more interestingg–I then knew what the *second* to last subject would be. That still left two. Chemical and Biological. Which one opened the division? I truly didn't want to down play any of them, or make you, the reader, feel that they were placed in order of importance and danger. That is not the case. Please keep in mind none of the topics are placed in an imperative order. So with the final two, after a while of deliberation, I decided. I flipped a coin. You are now about to learn about chemical weapons.

Chemical weapons, in my opinion, are the most scary prospect. After reading about the agents, you will see why. I am not a scientist. Unlike with biological weapons, the symptoms rarely mean anything as far as seeking help or treatments. With biological weapons, knowing the symptoms can increase your awareness, and in turn increase your chances of survival. With a majority of chemical weapons, the *majority* of the population will not have access to the proper antidotes. In short, chemical weapons kill. If you do not die immediately, your chances of survival can be slim, and those who do survive a chemical attack with a deliverance of a full dose, will be debilitated in some cases up to three months.

It's sad. It's the truth. Although it seems bleak, some chemical weapons allot a minimum exposure, and immediate symptoms that afford the victim time to seek decontamination.

Chemical Weapons

Chemical weapons can be basically placed into four different chemical agent categories.

Choking
Blister
Blood
Nerve

The deliverance of chemical weapons primarily are through explosives, however they can be distributed via an aerosol, smoke, fumes, droplets. They are silent. Unlike Biological weapons where effects can take days to surface, chemical agents and their effects appear within minutes to hours.

There are several names for the different types of chemical agents. I will list the most commonly known under the category in which they fall. Again, I am not a scientist, this is information I have learned through experts, studies, and research. I'll try my best to give you an understanding of each.

Choking Agents
Leading Agent-Phosgene

Choking agents were first used in 1915, and accounted for eighty percent of chemical weapons deaths in World War I. It is primarily delivered through a liquid fille shell that explodes and delivers a vapor. It is colorless, but extremely unstable, and lacks persistency. It's about four times thicker than normal air and works best to linger in low level areas such as trenches and holes. It carries a sweet smell similar to that of freshly cut grass, or hay. How a choking agent works is it attacks the lung, fills them with fluid and basically the victim dies from lack of air or, in an essence they drown in their own lung fluids.

For those who receive a lethal dose, choking agents deliver a slow, agonizing, death, in which the victim actually struggles every step of the way with every breath. Although, rare, in some cases the victim will only live two hours post exposure.

Exposure to a choking agent is signified at first by coughing, choking, a tight feeling in the chest, nausea, vomiting, and headache. This phase can last from two to twenty-four hours before progressing into a fatal pulmonary oedema. As the symptoms persist, sputum may be foamy, and patient may get pale and clammy. Death occurs within forty-eight hours. Keep in mind, the absence of symptoms, or abundance of symptoms at the onset of exposure has no bearing on a prognosis. Victims with little or no initial symptoms can still develop fatal pulmonary oedema, while those with severe onset symptoms may not develop lung injury at all. Those who make it past the forty-eight hour mark, usually recover.

There is no known antidote. Gas masks and protective clothing are the only deterrent. Nonetheless, since choking agents favor low level areas, high plains, may give some safeguard.

Blister Agents
>   Leading Agents-Mustard, Laced, Phosgene

>   Blister agents are used to debilitate more than to kill. Though they can cause fatalities. These agents burn and blister the skin, attack mucus membranes, blood producing organs, and the lungs. Respiratory damage occurs when inhaled, and we ingested, vomiting will occur. Different agents produce different effects.

Mustard Agents
>   These are oily based, and brownish, yellow in color. They carry a garlic or mustard smell. Not always are the effects prevalent, sometimes there is a delayed reaction up to an hour. Proper decontamination is recommended whenever there is a risk of exposure.
>   When in contact with the skin, a reddening occurs first, followed by a blistering and ulceration. Mucus membranes become destroyed when the agent comes in contact with the eyes. Death usually stems from inhaling the agent or ingesting it.
>   There is no antidote for this agent, protective clothing and a gasmask, plus full decontamination. Mustard is persistent in the air for thirty-six hours after release.

Lewisite
>   This agent is an oil based agent with no color. It is heavier than mustards. This agent not only causes blisters, it causes system breakdowns when inhaled. Pulmonary edema, diarrhea, nervousness, weakness. In large doses, death occurs in ten minutes.
>   Internal symptoms, like with the mustard agents, have a delayed appearance. However, unlike the mustard, Lewisite burns the skin immediately upon contact. This is an obvious warning to the victim, which would usually cause the person to seek decontamination. Immediate help lessens the severity of lesions.

Lesions are violent, blistery and can cause tissue death. Deep lesions show signs of gangrene. When inhaled, the nostrils will burn, and sneezing will occur similar to that of an allergic reaction. However, the more amounts inhaled, the more severe the symptoms become. Contact with the eyes can cause blindness.

There is an anti-dote to this agent, however, the antidote itself can be toxic. Antidote creams are available to ease the lesions. However, even with the anti-dote, decontamination and cream, lesions will still appear. If victim survives exposure, they should be watched closely for several weeks for relapses. This agent is still prevalent in the area of attack for up to one week, depending on the amount of agent used.

Phosgene

This agent is a skin irritant agent with an odor of sulphur. It causes extreme itching, then tissue and skin damage. Lesions are predominate. This agent is one of the most painful, striking the nerves as well. At first white spots appear on skin, then a redness. Skin soon blisters. Eye exposure will result in blindness. One of the scariest aspects of this agent is that from the skin itself, inhalation can occur causing pulmonary damage and in some cases hemorrhaging of the digestive track. Recovery can take up to three months.

Decontamination is the only means of treatment. There is no anti-dote for this agent.

Blood Agents

Leading Agents-Hydrocyanic Acid, Cyanic Chloride

These agents are least likely to be used because they are extremely unstable and tend to dissipate rapidly. Even though the United States kept a stock of blood agents, they are rarely even considered for use because they would fail to be toxic rather quickly. They work basically by causing tissue and organ deterioration through excessive carbon monoxide in the blood. This agent inhibits cells from taking oxygen to other parts of the body. This agent, in high doses, kills rapidly.

Upon exposure you probably would experience the longest six minutes of your life. First your breathing would be deep and irregular. After about thirty seconds you will start to violently convulse. Two minutes later you stop breathing, and then you slowly suffocate, no air inspiring into your body, until your heart stops.

There are effective anti-dotes including sodium nitrate. Gas masks and special clothing serve as a protection.

Nerve Agents

Leading Agents-Sarin, Tabun, Soman, GF, VX

Colorless and carrying a pleasant fruit smell, nerve gas is considered the top threat on the list of chemical weapons. They can be dispersed in many ways, and studies show that many countries have the capabilities to make nerve agents. Nerve agents were first introduced by the Germans in WWII. The US currently stockpiles, Sarin and VX.

Nerve Agents are pretty self explanatory by name. Your neuroglial system is needed to control everything in your body. Imagine, that neurological system being paralyzed. What would happen? A chain of events. Convulsions, paralysis, respiratory failure, cardiac arrest. All while your brain is still functioning and aware. Immediate decontamination and an antidote are recommended.

Death occurs rapidly, sometimes though taking up to an hour. VX and GF, are favorable weapons because they only take a minute for symptoms to start and death occurs fifteen minutes later.

There are antidotes, Atropine is the commonly known. However most Americans would not have atropine easily accessible within the time frame needed to save their lives.

Chemical Warfare Tidbits

To wear or not to wear a gas mask. Although you may smell, and see the gas, gas masks are useless unless they are used properly. Plus, as you have read, most agents absorb through the skin, so if you are not wearing special protective clothing, having the gasmask would be in vain.

Familiarize yourself with the colors and smells of these agents. That is your first line of defense.

Some indication that a chemical weapon has been used would be an abundance of dead animals, and insects. Low flying airplanes are a viable source of deliverance as well.

No gasmask? Will you die? Here's something that you can do. Flee as far away from the area as possible. Remember, unless large quantities are used, chemical agents are very localized. Find a convenience or meat store freezer, shut off air flow to the freezer and vents. This should provide some air protection while the levels of chemical lower on the air.

Chemical weapons are highly fallible. They will fail to carry the devastation if distributed wrong. They linger in one area and dissipate quickly. Rain, wind, snow, temperatures, all play a huge factor in the efficiency and potency of chemical weapons.

With chemical weapons, Mother nature is your friend. Nature breaks them down.

Biological Weapons

◆

That which nature creates is the deadliest of them all. Nature has a way of cleaning house every so often. Why man feels he must have a hand in that is beyond me. Yet, man does. He enhances what is already lethal. A virus, or bacterium, that is capable of bringing man close to extinction. History proves it. The Bubonic plague, the Spanish flu.

What the average person doesn't know is that Spanish Flu is a swine flu. Though Spanish Flu is considered eradicated, swine flu happily still exists. Bubonic plague didn't leave this earth centuries ago, it still claims at least two thousand lives every year.

One of the biggest differences between biological weapons and chemical agents is time. Biological weapons take time to incubate, develop into symptoms, then ravage the body. Time also allots time for a cure. Although there are biological weapons that have no vaccine, they do have means to combating them.

So now that you are breathing easier–no pun intended–don't. Biological weapons have no smell, you can't see them, and by the time you know that you have been hit, it may be too late to stop them. They linger in the air, and if enough people suffer from the illness, it very well may become the air, never to leave it.

Did you know that in 1918, the Army Surgeon General actually warned that if the flu continued on its current course, mankind was facing extinction?

What saved man? God? Medicine? In my opinion lack of superior air travel played a key role. The virus did not have enough time to circle the globe before it lost all strength.

But God's weapons, are not the same as man's weapons. Though basically the same makeup, biological weapons are created in the lab and based on what nature has already provided.

I will explain several weapons that would be commonly used in the event of a biological attack. Very basically, and very briefly. These descriptions can also be useful, should nature decide to release on of these bad boys on us. Again, as I have repeated mega times thus far in this book, I am not a scientist, nor a doctor. For further and deeper reading, research is always recommended.

The weapons are not listed in any order of importance, strength, or weapon viability.

Common Biological Weapons

The following are biological weapons that, more than likely, would be used in a biological attack.

Anthrax

Anthrax is an animal originated illness that can be transmitted through the hides of animals, it also breeds in the earth. There are three types of Anthrax: Inhalation. Cutaneous. Ingested.

The first of the three is the deadliest and rarest. Inhalation anthrax starts as a common cold, accompanied with a fever and progressively gets worse as the days move on. It is marked by a widened mediastinum strip, and a tightness of the chest. Basically, your lungs crush. Inhalation anthrax is ninety to a hundred percent fatal if left untreated. In order to get inhalation anthrax one must inhale at least 8,000 anthrax spores. A single spore is about a fifth of a hair strand. This dose would have to imbed itself deeply into the lungs and cultivate. For there to be an epidemic of inhalation anthrax, a large dosage would have to be propelled through an aerosol.

Cutaneous anthrax is the most common and least deadliest of the bunch. It has a twenty percent fatality rate if left untreated. To get this form of anthrax an open sore or open area of the skin would have to be exposed to anthrax spores. The spores get into the open area and infest. The sore then becomes a lesion that grows and gets black. Because of the lesion, usually prompt medical attention is sought.

Ingested anthrax begins like a case of the stomach flu. Cramping, weakness, dizziness, vomiting, diarrhea. It has a fatality rate of ten to twenty percent if left untreated. Ingested anthrax is caught when a person eats food that has the anthrax spores.

There is a vaccine for anthrax, but currently that is reserved for military personnel. All forms of anthrax are not contagious, carry an incubation period of one to seven days, and all forms of anthrax respond extremely well to antibiotics.

Author's note–during the time when I was compiling my notes to make this book, the nation and world were in a panic. With heightened terrorism acts, and a war on terrorism in progress, Anthrax was being released in selected locations. Now, keep in mind, by the time you read this, you should know the source, and outcome. But as of the moment as I write this, I must state, that *if* this was a biological act of war and not some insane criminal move, then the good people of this world should be grateful it was anthrax and not something else. Then again, by the time you read this, you may say, 'ha, they did release something else you moron'. Just keep in mind: Every year, in America, twenty-thousand people die because of the flu.

Botulinum Toxin

Botulinum toxins are a neurotransmitted disease. They affect the neurological system that controls muscles, respiration, the tongue and so forth. If delivered by air, the toxin could produce symptoms similar to food borne botulism. Which are atypical stomach flu symptoms.

After an incubation period of two to six days, inhaled botulinum begins with weakness, dizziness. It progresses to lack of salivation, causing the throat to get sore and swallowing difficulty. Crusts form on the mouth, muscle breakdown occurs, and it is followed by the inability to breath. Respiratory distress and respiratory paralysis is the main cause of death. Without ventilation and tracheotomy, fatality would be sixty percent.

There currently is an antitoxin for botulinum, but as of the writing of the book and last research, it was still under study.

Cholera

Caused by the vibrio cholera bacillus, people usually get this disease by consuming food, or drinking water containing the source. It is highly contagious and is excreted through the intestines and vomit of those infected. Many cases of cholera can be traced to improper handling of the dead, and the secretions of body fluids.

Cholera is marked by severe vomiting, and extreme amounts of watery stools. Diarrhea is so onerous that the patient will excrete themselves into dehydration. At that point there is shock, then death.

Cholera is more than likely to be placed in food or water in the event of a biological attack. Cholera can be treated with intravenous fluid replacement. There is a cholera vaccine that is regarded as having a 50% effective rate. This is not readily available to the general population.

Hemorrhagic Fever (HF)

More commonly known, the Ebola virus, is a form of hemorrhagic fever. A lesser known, and just as contagious HF, is Congo-Crimean. Hemorrhagic fevers are highly contagious, through contact with infected person's bodily fluids. The most common form of distribution would be through the water supply. At this time, it is not known to me, and I have yet to find out, whether or not there is an aerosol form.

Ebola has a fatality rate of ninety percent if left untreated. CCHF, has a fatality rate of thirty percent. There is no known vaccine for these, but treatment varies. Isolation of patient is recommended, and safe handling extremely encouraged. The death rate is still high, even in treated cases.

Basically after a two day to three week incubation period, a fever sets in. Patient exhibits flu-like symptoms that develop in severity. Vomiting, bloody diarrhea. Fever can become so pronounced that internal organs begin to liquify. It received its name because of marked internal bleeding. It is not uncommon for a patient in the latent fatal stages to bleed from every body orifice. Eyes, nose, ears, rectum, pores.

In primitive medical areas, death to those with HF is imminent.

Small Pox

Considered eradicated by the late seventies, small pox is considered one of the most prevalent forms of biological weapons. It still exists. It is still out there. The disease has an incubation period of seven to fifteen days. It starts out with fatigue, body aches, headache. It progresses into eruptions of the skin. Pox marks, that eventually become pustules that erupt, scab and eventually heal. They leave large, deep scars. Untreated small pox carries a fatality rate of thirty-five percent. In those recovering, joint deformities and blindness is common.

A good bit of the population was inoculated against smallpox up until 1972. After that, mostly the vaccination was a choice, and it eventually went from public access by the end of the seventies. Experts are in debate on whether the vaccination is still effective, however, an old inoculation is better than no inoculation. There is a small pox vaccine. At the time of this book's writing, a bill is being moved through legislation to innoculate every single American by the end of 2003. In an essence, and more than likely, eventually re-adding small pox as a required vaccine for all school age children. This author feels, the inoculations should have ever stopped. Nothing can be considered eradicated from this earth when it still exists in laboratories.

Plague

Probably my favorite because plague is more than a biological weapon. It is an everyday fact of life in this world. If a pandemic would break out of plague, without the help of man's weapon, we very well could lose over half the world's population…at least.

Plague is pretty interesting. It stems from the Yersinia Pestis, a flea. It bites the rodents, rodents tromp through the food. The flea also bites people and that is why this natural disease was so out of control. However, it is rare to see plague before the month of March, and after June. Those are the months that Yersinia Pestis are prevalent. So, of course, if a major pandemic of plague hits around December, we would know something out of nature's control is up.

Basically there are two types of plague. Bubonic Plague strikes the lymph nodes. It has a two to ten day incubation period and starts with your typical fever, tender nodes and gradually strike the central nervous system. Bubonic Plague was also known as the black plague because its eventual turn to septicemia, the poisoning of the blood. Internal organs melt, and seep from body cavities in a black, bloody manner. It is very painful and if not treated is almost always fatal. Of course if treated, the mortality rate is fifty percent.

There is also pneumonic plague. Which attacks the lungs. It has the same incubation period, and begins with fever and cold symptoms, gradually digressing to a biblical size case of pneumonia. The patient eventually can't breath, can't produce a productive enough cough to eject sputum, and they die. It isn't as visually ugly as Bubonic plague, but it's mortality rate of a hundred percent is nearly irreversible with treatment or without.

Little known to people is that there is a plague vaccine. No, I am not joking. It is not widely available but it has a high effective rate against the effects of the Yersinia Pestis. A booster is recommended every two years.

Lesser Known Biological Weapons

I really could have left these few out. But why? Though the chances of an enemy using them are slim, they are viable biological weapons. But more importantly than that, they, on their own, without any help from man, are extinction level diseases. I find them very interesting. See how many you have heard of:

Rift Valley Fever-This is a virus, not a bacterium, and is caused by mosquitos. Ever hear of it? We are all familiar with Pink Eye, right? Well that is how this one starts out. Conjunctivitis sets it, and despite attempts with medication, it progresses to abdominal tenderness. Then you feel better. But wait. The symptoms hit you all over again, grow increasingly more severe. A small percentage of those infected develop hemorrhagic fever and die. The average mortality rate for Rift Valley Fever is fifty percent. Those who recover from the disease are usually blind, or have lost all focal ability.

Clostridium Perfringens Toxins-Sometimes known as gas gangrenes. These toxins carry a large amount of CO_2, which cause intense swelling and pressure on internal organs and tissues. These toxins, can be delivered via aerosol, skin, or ingesting. In any case, this toxin injects a bacteria poison into the bloodstream, mimicking the devastating effects of gangrene in just a few hours. Antibiotics are effective, but it must be caught in time, and quickly enough. If a level of toxins is high in the blood, the mortality rate is near a hundred percent.

Ricin-this comes from the seed of the castor plant, and is an accessible means to a biological weapon. No one knows of this, which is terrible because it can appear anywhere. How it works is, it takes over the RNA and kills the cells. Ricin is a white, non specifically shaped substance. If ingested Ricin takes on the appearance of severe food poisoning. If inhaled, the symptoms look exactly like pneumonia. It is all natural and hard to detect. There is no anti-toxin, no cure, and is almost always fatal.

Nuclear War

◆

Maybe it was my mother, I don't know. That complex gene she placed in me. A fetal maternal, paranoia transfusion. Whatever the case, I learned about nuclear war. I don't recall my mother really ever bringing up nuclear weapons. My father on the other hand did. He knew my obsession, he knew my fear of nuclear weapons. He actually purchased my first survival book for me just to shut me up. I'll always recall his words of wisdom when I would ask him, even as a young adult, if he thought there would be a nuclear war. His response was always the same, 'If they use the weapons they won't use that many. We feed seventy-five percent of the world's population. They won't bite the hand that feeds them. They may poison that hand, never nuke it.'

Inadvertently he'd switch my mind set to biological weapons. Plagues, biological weapons, chemicals, for some odd reason, they never scared me. But nuclear war did. So I learned what I feared. I studied it, and still do. Searching for information that I don't know. I taught my children too, and still do. It surprised me recently when a friend of mine's child had no idea what a nuclear weapon was. When I explained, he said, 'oh, yeah, I heard of that.'

Heard of that? He was fifteen years old. When I was fifteen years old I was searching out viable bomb shelters. Nuclear war was such a massive threat years ago, that it was actually forced down our throats. Today, it is considered a minimal threat, and our young people barely know what it is. It is up to us to educate them. Teach them.

The truth is, at the time of this books writing, despite any disarmament, twenty-thousand nuclear weapons still exist under control of the Untied States. That is only the Unites States. Third world countries have the bomb. A recipe for making one was readily available for years. The Former Soviet Union has multitudes of suitcase size nuclear weapons that are unaccounted for. So how can we with a clear conscience even think that there isn't a threat. With terrorism so high, it is a bigger threat than ever before. And even more deadly. Granted their capabilities are limited; any bomb that they have would be at the very least fifty times less the strength of one of ours. But they have one thing in their favor that makes them the ultimate threat. The element of surprise. If China sends a dozen nukes this way, we'll see them coming. We can try to take them out, if not, there is still a warning to try to run for cover. But one suitcase size nuclear weapon in the trunk of one car. Boom. No siren. No warning. No time.

It is the intention of this division to give you some facts about surviving the effects of a nuclear war, and how you can protect yourself and your family. Helping to know the facts, will help you to stay alive.

Warnings

Barring a nuclear car bomb device, or a bomb being dropped by a low flying small plane, there would be warning time. If you are on top of the news, and have watched the national situations digress, what is called a 'strategical warning' would be given by the news media, government sources, etc. These sources will alert those that a nuclear attack is a high probability. Back in the seventies, one would have watched for the Soviet Union to evacuate, that was a tell-tale sign back then that they were preparing to launch. This, giving time to those in high risk areas to improve shelter, or evacuate.

Tactical warnings are given when our military receives information that a nuclear device has been launched. After confirmation, and reconfirmation, an alert is sent out and then relayed through the National Broadcasting system.

In the event of an all-out nuclear strike, the first weapons to hit our soils would give ample warning, at least, to the general public to seek shelter. SLBM missiles would be fired at designated military targets, and would be fired from submarines. These strikes would set off explosions heard all over the country. The sky would light up with bright vivid flashes, and a loss of power would occur. This happens approximately 15 minutes before the intercontinental ballistic missiles (ICBM) arrive. These are the ones that target our cities.

Make no error in judgement. Should you be out and you see the bright flashes of light, experience electrical power loss, don't think to yourself you still have fifteen minutes. Blast distances are hard to judge. Be safe. Take cover and wait at least two minutes post the flashes to ensure that a blast is not heading your way. After that, seek a stable shelter.

There are four 'definites' in a nuclear explosion. Meaning, four things, that without a doubt will happen. A warning, is not one of them. As you know, there are ways to deliver a nuclear attack without a warning. The four items are explained below, and they are: Electro Magnetic Pulses (EMP), the blast, fallout, radiation.

Electro Magnetic Pulse Effect (EMP)

We have all heard quite a bit about it, though many of us may not know what the proper name for it is. It is the action that occurs when a nuclear weapon is blasted at a high altitude. The EMP effect disrupts all electrical devices. In an essence, anything that is running, phones, cars, radios, televisions, will cease to run the moment the bombs are burst.

Surface or sub-surface burst would cause the same effect but more so in a close range area. A nuclear weapon air burst at an extremely high altitude, would disrupt power for hundreds of miles. One bomb. In the technological world we live in today, realistically, another country could send us into total disarray and panic, not by blowing us up, but simply strategically exploding several high altitude nuclear weapons at the same time. Imagine, all power lost in the United States. All power. We would collapse, and that is a scary thought. There are means to protect from the EMP effect, most government installations have it. But the majority of businesses, homes, and so forth would be majorly effected.

It truly hasn't been proven, and it's only in theory, but it is said, that anything powered down by the EMP effect will never work again. Though there are some who say, old cars would be able to restart because they do not rely on electrical sparks to get them going.

Blasts

There are three types of nuclear blasts. Air bursts, surface bursts and sub-surface bursts. The later of the three would be the case in an explosion taking place in the ocean, or basement of a building. The most common would be air bursts. This is when the nuclear bomb detonates about a hundred or so feet from the group. This, especially in hilly areas would produce a larger scale blast effect. Winds of destruction, fireball would be greater because the earth would not be an inhibitor. There would still be a crater, but not one as large as a surface burst. A surface burst is exactly how it sounds, the bomb will not detonate until it hits the ground. This type of explosion would cause a wall of dirt to mix with the destructive winds and fire. The earth would decrease the fireball's destruction radius.

All three explosions, if all are of the same size bomb, would have the same amount of radiation release. However, sub surface, and surface explosions would limit the range of the initial radiation dose. Air bursts end that first dose further.

Aside from the obvious blast burn, there are two other things that could occur immediately prior and during the explosion. Flash burns, and flash blindness are a high risk potential, even when you are not close to ground zero. These are caused by the bright light that happens at the onset of detonation. Blindness is imminent if you look directly at the flash.

Fallout

Beta Particles.

Beta particles are high speed electrons produced by the explosion of the atom. They are radioactive material which mix with the ash of the destruction. Usually they are thick and ashy in nature. These are not to be confused with radiation, though they are radioactive. They look like a white-grey snow and arrive in what most would call 'fallout'. Fallout is the ash, dirt, and debris, minuscule and dusty in size, that is too heavy to float into the atmosphere, so hence it 'falls' quickly to the ground, within minutes, after a nuclear blast.

Without being scientific, here are the facts. You cannot see the radioactive material. It will exist, from the initial blast in the ash, debris, and dirt that sprays about. They cannot penetrate concrete, nor can they penetrate typical clothing…At first.

Buildings that would protect from fallout, protect from radioactive fallout as well. Your clothes will suffice, but beta particles would burn their way through any thin clothing after about an hour. Particles left on the skin will cause burns that take a long time to heal, remain open wounds, and more than likely become ulcerated.

Winds can cause beta particles to get into eyes and nostrils. These particles are not what delivers the deadly dose of radiation. Unless, of course, you are wandering in the fallout and ash. Thus, causing continuos exposure to the particles. Or, if you eat or drink contaminated food and water with fallout particles. The ingested means will result in radiation poisoning.

The particles are not 'hot' to the feel, so therefore when they land on you, they don't sizzle like a hot poker. However, after about fifteen minutes, you will feel your skin begin to burn. Wash the area immediately. This will rinse off any other beta particles that remain, but will not stop the burn mark that will appear within a few hours.

Treat the burns by keeping them clean, and with antibiotic ointment. As mentioned above, they will be open sores and hard to keep clear of infection. For a true example of a beta particle incident, see the 'Whoops' story at the end of this division.

Radiation

Radiation is by far the thing to fear in the event of a nuclear war. Mainly because you can't see it. Dosages of radiation are measured in what are called roentgens, or quite simply abbreviated, 'R'. Each roentgen is a unit, and the human body can only take so many units in a lifetime. The units delivered are measured as units delivered per hour. For example, if the radiation level is 100r per hour, that means for ever hour a person is outside they are receiving a hundred units of radiation.

It is said that a person can receive a dose up to 100R, with little or no side effects. If the dose is received at one time, slight nauseousness may appear. Other than that, there would be no long term effect, or grave illness. No effects would likely to be felt if a total dose of 100 R were received over several days or weeks.

A dose of 200R would cause illness in ninety percent of the people who receive it. Nausea, diarrhea, skin discoloring, some hair loss. A complete recovery would be expected and causalities of this dose would only be experienced under extreme circumstances. However, if the dose was received over several weeks or months, the effects would be lessened or mild.

It is said that half of all those who receive a dose of 450R would die. Those who survived would experience symptoms to those listed above, but more in a serious manner. Skin may peel, discolor, flake off.

Any person or persons receiving a dose of 600R or more would die within two to three weeks. The suffering would be great. Typical radiation symptoms compiled with hemorrhaging.

If an initial dose of radiation, or extremely large doses of 800R or more occurred, such as would be received by an individual who was outdoors the first hour and a half after the explosion, that individual would be dead within a day or two.

Keep in mind, it is far more dangerous for the human body to receive higher doses all at one as opposed to a total dose over several days or weeks. Exposure over several weeks and even months can allot the human body time to repair cell damage.

Radiation targets the thyroid gland, hence the reason for long-term effects. Also, older individuals have built up some resistance to radiation poisoning. The older a person, the less likely they will be to show symptoms.

All individuals should expect some exposure to radiation. Especially those who assist in reinstating law and order. Keeping exposure minimal will reduce symptoms and long-term side effects. Also, taking Potassium of Iodide fights off the effects of radiation poisoning. A funny, thing is, The Former Soviet Union had such an extensive evacuation plan, that their shelters and citizens were given potassium of iodide a head of time.

So how do you do it? How do you survive the radiation phase?

First, keep in mind, the levels will drop to safe levels within time. In the event of all-out war and an all-out attack, those in a shelter can safely assume that after two weeks, radiation levels dropped to somewhere around 2R an hour. A bit more the closer you get to ground zero.

If for example, two fifty-megaton warheads, or one 5 KT warhead was dropped on your city, the initial radiation dose would be about 2000R per hour. Weather conditions may hinder or help, but typically, the levels would drop after about eight hours to around 200R an hour. Forty eight hours, it will drop again. More like to about 60R per hour possibly more. From there on, the levels take longer to drop. These levels are why it is imperative that you stay in shelter, stay protected and do not emerge unless absolutely necessary. The longer you stay below, and in safe shelter, the more chance you have of surviving any radiation you may receive.

Radiation at mega levels begin about one half hour after the explosion. This is due to the fact that the radiation particles are shot up into the atmosphere and they take time to come down in gamma rays. A half hour, then it comes.

But what in the case of a small nuclear weapon? Laymen made, terrorist delivered. These weapons typically have about one fifteenth the strength of the bomb dropped on Hiroshima. Though there are rumors that there are bombs with as much strength as the Hiroshima bomb. Do not let the lesser size of explosions, fool you. A detonated suitcase size nuke would take out about three city blocks. Flatten it. But the radiation levels would still be there and would still fan out, spreading about the areas near and surrounding it.

The radiation levels would be less, but they still would be at dangerous levels. Take shelter, cover, and stay there.

Radiation dosage meters, or Geiger counters can be purchased. You can also make an effected dosage meter at home for about a forty dollar cost. I do not get into directions, but the book, '*Nuclear War Survival Skills*' which is also readily available on the net at this time, tells in full detail, exactly how to make a dosage meter. This is well worth looking into.

Radiation poisoning and consumption radiation poisoning are a huge threat because stupidity will cause mas casualties. In today's society, many do not know, or care to know. An ounce of knowledge, in this situation, surely is worth a pound of cure.

Points to Remember
I like to think of these little sections that I add at the end of each division, as sort of cliff notes. Nothing is written in long, detailed explanation, more of a common sense deliverance. Some of the suggestions may have been mentioned earlier, but it never hurts to reiterate. These are some final tidbits to help you out.

Always remember not to drink any water, or eat any food that has been exposed to radiation, or has had fallout particles on them. Water that stands a chance of being radioactive, should be fully decontaminated prior to drinking. (See section on water)

Remember to shield your eyes when you are outside, immediately following a blast, and post your emergence from the shelter. Eyes can be damaged by radiation. Dark sunglass with a high UV protection will work.

If you are outdoors when the blast occurs, and fallout starts to land, cover yourself with a thick blanket and shield your eyes for protection from beta particles.

Something that a lot of people are not aware of. We worry about the safety of our kids in school during an attack. After 1950, most schools being built were built with a bunker style center. If you go to what would be the center of any school building constructed after 1950, most will have a center that has no visible windows. This center is the place in the building, with the exception of the basement, that is the furthest point from glass windows. It is designated to be able to move the children to this location in the event of an attack if they cannot reach the basement.

In a crowded, cramped shelter, Psychosomatic symptoms will appear, and carry the same symptoms as radiation poisoning.

If a lot of people in a small cramped shelter are suffering from beta burns, try your best to keep everyone calm, and wound dry and free from debris. Infection will spread and beta burns are open sores.

Remember that if you see a bright flash take cover wherever you are. Wait about two minutes, to make sure it is not a close range explosion, before leaving your cover to seek better shelter. The flash may be closer than you think. People in Pittsburgh may experience the 'boom' and 'flash' seven minutes after a nuclear device is detonated in New York. Wait two minutes, if no explosion, seek shelter.

If you are in some sort of shelter. Do not leave when the warning comes if there is even the slightest chance you won't make it to your shelter. Seek shelter close by. If you are outside, find a ditch. First comes the radioactive fallout, then comes the radiation. What is too heavy to go up into the atmosphere comes down first. What makes it into the atmosphere is the most dangerous, and that does not begin its decent for a half an hour. You have a safety window post the blast between fallout and radiation to seek stable and safe shelter. Anything post the twenty minutes is putting yourself at extremely high risk.

Finally, don't be stupid. Stupidity in these cases will kill. I have given the top stupidity move a name, it's the, 'I Know I Can Make It Home', syndrom. Most men, if not ALL men who have a wife or family at home will fall into this category. As a man reading this, with a family, you know what I'm talking about. First the blast, you survive, then the fallout. You know approximately where you are, and how long it would take you to make it home on foot. You think "I can make it. I have enough time. I have to get to my family." Wrong. The best thing you can do for your family is to stay put. Again, stay put. Wait at least forty-eight hours, the levels will drop enough for you to leave for twenty minutes or a half an hour without being disabled. If you get caught in that first hour after a hit, you will die within a day or two. Ten minutes in that amount of radiation will kill you. A prepared family will know not to expect you for days. If you have the contingency plan, they know to wait. You are fallible, and your family needs you alive and well. They don't need you to die, or to be a sick burden in the shelter. To everyone, think of your loved ones, and stay put.

Whoops

By far one of my favorite stories. I could tell it over and over again, that's probably why it made it into this book. This story is completely and one hundred percent true. Though I tell it in my style, it is not a 'Jackie' tale. Kept secret for decades, this story is so worth mentioning, and so deserving of the title I have given it of, *Whoops*.

In March 1954, the United States Government decided to test the new H-Bomb. They called the test 'Bravo'. This new bomb was supposed to be a five-megaton nuclear weapon, or five times larger than the one dropped on Hiroshima. They picked a quaint ocean spot about a hundred and thirty miles from the Marshall Islands. What ended up happening, was the bomb turned out to be at least fifteen megatons, and unseen by the government meteorologists, a wind shift occurred.

First the explosion was massive, reaching so far into the ocean, scientific observers of the bomb stated the white dust that fell could only be coral reef. One scientist had reported the flash kept growing and growing, he thought the sky was exploding.

Bomb tested, World didn't end. Sky didn't blow up.

Whew.

About a half an hour later, that 'white stuff' fell on a fishing boat about forty miles away. Twenty-three fishermen were on that boat and the ash floated down on them. They kept fishing. Interviews with the fisherman state they heard the boom, but kept on working. But that's not all. An hour and a half later, that same white material, now known as fallout and beta particles, started falling like snow on the beaches of the Marshall Islands. Well, the islanders didn't think anything of it. After all, they heard the blast, felt the ground shake, to them the dust was ash from a nearby volcano. They moved about their lives, blew the ash off of surfaces, drank the water, and ate the food. The children even played in the fallout as if it were snow. Happily they did, barefoot in all.

In a bunker on the island, soldiers doing a typical readout after the bomb, saw the reading skyrocket. They immediately notified US authorities who ran the test, and the US, without hesitation, removed the soldiers from the island.

Oddly enough, they waited an entire day before getting the natives from the island. When they did, by the time US officials showed up, the natives were vomiting, losing their hair and the skin was literally peeling from the body.

To those who did the Bravo test, the mishap was tragic, but beneficial because they inadvertently received results on the effects of radiation.

Now I'll share with you my writer theory on this. I think a batch of those testing guys sat around wishing they could get accurate results on the effects of radiation. But ethically they couldn't knowingly expose people to it. So they picked a spot, preplanned meteorological igno-rance, dropped the bomb and gathered the data.

After saying, 'whoops', in 1955 the US government paid two million dollars to replace that fishing boat that was first hit with beta particles, and as compensation to the twenty-three men who became ill from radiation. The simple testing is still causing financial drainage to the US government. As of the writing of this book, thus far, they have paid out over Bravo testing in compensation and damages, sixty-eight million dollars.

I love that story.

God's End

All right, we can face it. Unless you are lucky enough to be one of the 144,000 marked by God, as stated in the Bible, then being survival savvy is pretty much useless. For the most part we are all screwed if God Himself, tosses in the towel. So why did I include it? Well, I think the entire story from the Book of Revelation is cool. And, it is the ultimate apocalyptic scenario.

Two of my novels deal with the signs from God's end perspective, and in writing those novels I had to do extensive research into the Bible. Let me tell you, I read it. I learned it, studied the Book of Revelation, and theorized my own interpretations. That is what you'll read in this section, my theories on what the signs could be. The actual signs are taken straight from the Bible.

Warning–I'm probably about to offend a few of you with the following three paragraphs, so you may want to skip them if you get easily offended.

When it comes to The Bible, you really have to take it at face value. Even the 'easy to read' modernized lingo versions are often, at points in The Bible, a challenge. The rambling sentences, many times repetitive and poorly written passages, make it difficult to read, let alone understand. Do I believe it in? Yes and no. I believe in the premises and the basis for the stories, but like the fisherman who caught the twenty-foot trout, in my opinion, the Bible is exaggerated. I mean, the proportions of the walls in the New Jerusalem alone are enough to make you question the entire Book of Revelation. Most people when they think of Revelation, do not think of the walls' proportion or details like that. They think, fire from the sky, real hell on earth, sort of stuff.

The Book of Revelation, 21:16-17, states that the New Jerusalem is a cube. The walls as high as they are wide. The length, width, and height of the walls will be 1,400 miles. The thickness of the walls, 216 feet. Back it up. A wall 216 feet thick and 1,400 miles high? Hello. Let us not forget, that the author, John, also mentions in Revelation 21:17, that the angel used standard human measurement.

Barring the size of the wall that threw me into the doubt wheel, there's the simple fact that John wrote Revelation while in exile for believing in Christ. Thrown in prison. I do not doubt an angel came to him and told him about the end of the world, but I do think that John may have been a little bitter and hostile about his exiling and perhaps intensified a bit what he was told.

OK, now that the Jackie view is out of the way, let's continue on. The Bible states that only 144,000 will be saved to start things over again, to live within those huge walls of the New Jerusalem. Figures and numbers may not be accurate, biblical scholars often accede to that. John could have been told ten percent of the population, which would have made complete sense for the 144,000 figure. So by today's population, if something like the ten percent thing is true, then a lot more of us get a chance to live.

Revelation–after the visiting angel and explanation of why the angel chose John–then goes on to basically tell how the world will end. According to God; delivered in three forms. Seven Seals. Seven Trumpets. Seven Plagues. Those are what I am going to approach in this division. I will list what the Bible states each of those 'sevens' are, and then give my theory and interpretation on them, if applicable. Remember, they are my theories, so take those interpretations with a grain of salt. Here we go.

The Seven Seals

The seals (a wax circle used to keep an envelope closed) will be broken. Seven of them will initiate the start to the devastating end of the world. God's signs that we're all pretty much on our way to being toast. These are the plagues started by the breaking of the seals.

The arrival of the white horse. The bible states that this horse rides in to ensure victory. I see the white horse as a sign of peace. A peacemaker, or someone who initiates world peace. Better yet, what about an ambassador who finally brings peace to the middle east?

The arrival of the red horse. Revelation states this means war. This can be the unity of Russia and China against a foreign enemy. Or a war between the two red countries, Russia and China.

The arrival of the black horse in Revelation is the third seal, and explained that it brings justice. No waste of time. I see the black horse as a person. Someone who has been persecuted wrongly, and in return begins to stand up for the rights of others, becoming a very influential person. Martin Luther King was the black horse, in my mother's doomsday time.

The pale horse rides in next bringing famine and disease. This wipes out twenty-five percent of the population. I don't believe this prediction to be a worldwide twenty-five percent. Reason being that if you add up all the casualties predicted by The Bible, it ends up being more than they had people at that time. Yes, there is room for population growth, but I do not think the prophets of The Bible foresaw an increase in people like history really had. They had no census, so they had no way of knowing how many people there really were in the world. Plus, they still thought the world was flat. So, getting back. This twenty-five percent can be on bad summer in Africa.

Martyrs cry out for justice and swiftly strike back. That is the next seal to be broken. This was an easy one. The Islamic nations rising up. The Palestinians against the Jewish state.

A great earthquake, the moon will turn blood red, the sun black, the stars fall from the sky. Mountains and Islands disappear. Something so simple, it could easily be a seal missed. How, you ask? Think of a hot summer night. Night meaning, the sun is black. How many of those summer nights, have you seen the mood red? I can recall a lot. A shooting star or two, or a small meteor crashing down onto a minuscule island or nearby would cause a tidal wave that could conceivably wipe out that island.

The final seals predicts that fire will be thrown from the altar, horrible thunder and lightening, and great earthquakes. Take notice now how many times we are threatened in The Bible with earthquakes and fire from the sky. What is an altar? A place held in high regard at a place of worship. What about a holy place, where a volcano erupts. Earthquakes always accompany eruptions, and large quantities of smoke and ash would cause the storms to brew.

The seven seals were described in The Bible very simply. Deep and dark details were not given, the events were seemingly minimized to the extent they were preparing the reader for the building climax. The seven plagues, and seven bowls get a bit more into a graphic and gory detail, which you know, I will share.

The Seven Trumpets

'Then the seven angels with the seven trumpets prepared to blow their mighty blasts.' Revelation, 8:6. What a keen introduction to that chapter of The Bible. These are events that are to occur each time one of the seven angels blows a trumpet. They are pretty vivid and nasty; described in gory details at some points. John pretty much takes care of most of the explanations, but at parts where I feel I have a viable theory, do not fret, I interject.

The first trumpet blows and hail and fire mixed with blood, no less, is thrown from the sky. One third of the earth is set afire. A third of all trees and grass are burned. A lot of meteors consist of gas and ice, and when they enter our atmosphere, burn. Hence the 'red blood' portion. If impacted in the right place, a meteor a mile wide would cause this sort of damage.

When the second angel blows his trumpet, we're to expect a great mountain of fire being thrown into the sea. Volcano maybe? This will supposedly kill one third of all sea life, and a third of all ships.

Speaking of 'thirds' here blows the third trumpet. A flaming star called bitterness will fall from the sky into the ocean. Again, we deal with things falling from the sky. This star will turn the waters bitter, and many people are going to die from drinking the bitter water. Check this out, suppose some satellite covered in an organism from outer space drops in the Hoover Dam. It poisons the water. Hey, it's possible, and more original that me saying another meteor.

The fourth trumpet is supposed to bring darkness to one third of the sun, moon and stars. This will occur for a third of a day, and a third of a night. Basically, a huge eclipse will toss us into darkness for sixteen hours.

The fifth trumpet is one of my preferred plagues in the Book of Revelation. The locusts. The ground will open up, a black smoke will emerge, and a thick cloud of locusts will attack those who do not have the seal of God on the foreheads. Now, The Bible does not stop there. This little section, Revelation 9:3-10, really gets into details. People will be stung, but will not die from the locust bite, instead, painfully thy will suffer for five months. The locusts, according to John, are not your normal locusts. They arrive in a suit of armor and their wings roar. They have crowns on their heads, human faces, long hair like a woman, and teeth like a lion. Now that's some super locust. But…what if they aren't locusts as in insects? What do locusts do? Take all they can, devour all they can, and move on. What if calling them locusts are symbolizations? For aliens maybe? OK, just a thought. The sixth trumpet brings the arrival of the four horsemen. They come wearing armor of red, blue and yellow. Their mouths breathe fire that smells like sulphur and these four horsemen will cause the deaths of two hundred million people. A lot of people have likened this to nuclear weapons, me, I prefer solar flares.

The seventh and final trumpet is said to begin the third terror. It really isn't all that much, nor that difficult to associate with something in these times. Blunt and to the point, the heavens open up, lightening flashes thunder roars, a great hailstorm will cascade upon earth, and the world will experience a mighty earthquake. The third earthquake thus far, predicted to occur.

The Seven Plagues

Prior to the last blowing trumpet, The Bible introduces us to the two witnesses. They are chosen by God to witness all the destruction and not be harmed by it. Then we have the woman who runs from a dragon in the woods, before giving birth. Then the appearance of the anti-Christ, called The Beast. After he shows up, the seven bowls of plagues are spilled upon the earth. These seven are the real destructive ones, meant to pretty much bring the world to its knees. At least those who were worshiping the Anti-Christ. Of all the 'sevens' it surprises me that the finale isn't written in deep detail. One would think it would be. It is almost as if John was wearing down on things to write, and just jotted thoughts. They are spilled right before the chapter on the great prostitute, which has no bearing, but I had to mention it anyway. The seven plagues.

The first plague poured unto man is the plague of 'horrible malignant sores'. They break out on everyone who has the mark of the beast. I always thought this one to be an easy interpretation. Malignant sore could be smallpox delivered as a biological weapon onto the enemy in a religious war. The deliverers believe the 'enemy' to be evil and the beast.

The second plague is poured into the sea. The water becomes like the blood of a corpse and everything in that sea dies. Again, I do not think ths was intended for all seas. Back then they hadn't a clue how many oceans were to sail, seas to discover. A simple oil spill would also constitute the same effect.

Considering what history has told us of attacks on modern Jerusalem, this plague, the third can be viewed as viably happening. This plague deals with the killing of the murderers of God's Holy People. The rivers turn to blood, they drink the blood and die. If I read of a successful chemical attack on a water supply between two Mid-Eastern countries, I would actually contemplate if it was the third plague.

From the sun comes a blast of heat that burns everyone who is around. The fourth plague tells of this and how it forces people to repent. Though it is once again, fire from the sky, it is the most direct of them all. I see no other logical theory for this one other than a solar flare.

On the pouring of the fifth, those who worshiped 'the beast' are plunged into darkness and their bodies are covered with sores and pain. I've always tried to creatively find another answer to this one, other than eclipse. Maybe the dust cloud from a plunged meteor previously delivered causes a nuclear winter, and those without sun, could experience dead skin cells that cause their skin to peel off.

The sixth plague tells of the final battle. Good versus evil. With the drying up of the Euphrates River, the uprising begins.

The seventh and final plague is poured onto earth with it comes the greatest of earthquakes the world has ever experienced. Babylon crumbles. Cities fall. Islands are gone. Mountains crumble. Thunder, lightening, and seventy-five pound hail stones will fall from the sky down onto everyone. The world begins its next phase before Judgement day.

I really hope all of you enjoyed this section. Though it reeks of my sarcasm and cynicism at times, the biblical information is taken from The Bible. I had fun writing this section. I shared a good part of it with a close friend of mine. He asked how long it had taken me to compile this entire book from my notes and I informed him eight days. He merely let out a 'hmm', paused and stated, 'You do realize if you were more Godly, you could have finished the book in six and rested on the seventh.'

I found his comment enjoyable.

COMMON KNOWLEDGE

◆

This final section is more of a wrap-up, and reiteration of a lot of things covered in this book. This is where your knowledge is tested and perhaps, your own final questions are answered.

Frequently Confused Points

◆

Though not many, I have compiled a list of questions and statements that I frequently hear. Some are common misconceptions. I take a few moments to try to clear up some of the ones I hear often. Again, like most of the book, are listed in no particular order.

What if I know I am close to home after a nuclear explosion, and know how long it takes me to run home, would I be safe?

Why chance it? Are you certain your perception of surroundings is that great? If everything is tumbled, or in complete ruins, will you have your directional bearings? If you are truly certain it will take you little time to run home, then wait a couple days for radiation to drop enough that should you run into obstacles, the radiation dose won't kill you.

If we run out of water, and we find a store with bottled water, can we drink that?

No. The bottles are usually clear plastic, which allows for light and gamma rays to seep in. If this is your only option, you must follow the steps to safely removing radiation.

Can I use plywood, or a thick artist canvas to block out my basement windows?

You can, but not a safe means to keeping out radiation. A useful way to determine if your window barricade idea will help keep out radiation is this: Turn a blow dryer on full heat and air, place it close or flush to the barricade you have in mind (Foam, wood, clothes). By placing your hand on the other side of the barricade, check how long it takes for you to feel the warmth or air. Thickness and absorbency is the issue as much as blocking out the light.

I have received an Anthrax Vaccine, and Smallpox inoculation. So I'm safe, right?

If those two are the only viral strains dropped. Chances are, in an all-out, biological attack, one strain, even two will not be used. More like three or four different strains would be distributed.

Biological and chemical weapons were designed to wipe man into extinction.

No. They can be used for such, and more likely if wiping man into extinction, or performing genocide is the primary goal, then the deliverance of these weapons would likely be done by a single source or terrorist group. Biological and chemical weapons were designed not for the population, but for use on enemy troops in order to disable them.

If they drop a chemical weapon in California, the eastern winds will carry it over the country and wipe out the world.

Maniacal laughter enters here. A chemical weapon is just that, a chemical. Most had a dissipation factor of ten minutes to two weeks tops. After that normal climate changes and elements will break down the effectiveness of the weapon.

What are high risk, and low risk areas?

In the event of any strike, high risk areas are military installations, areas around them. Areas around missile silos. Larger industrial cities. Shipping ports. Some areas that are low risk for nuclear strikes are: Northen Minnesota, Central West Virginia, Wyoming, Western Colorado, North Eastern Nevada, also parts of Washington and Oregon.

Why should I stockpile food and ration my food, won't the government step in?

There's always that chance, but why would you chance it? What if the country suffers complete and total breakdown? What if those who are designated to emerge and distribute aid, suddenly go AWOL. Take no chances, get prepared.

I heard the term 'nuclear winter', what is it?

A nuclear winter occurs when so much dust, dirt, and debris are thrown into the atmosphere that it forms a blockade prohibiting the sun to peek through. Although not plunged into complete darkness, the light level will be low and plant life along with human life will freeze to death.

What is survivor syndrom?

It is when those who survive a holocaust lose all mental balance and become a physical threat to everyone around them. They become threatened easily, and set off without warning. Personality changes, twitches, and sudden outburst are preliminary warnings. Those who suffer from this theoretically will not recover and can be extremely dangerous.

Brand New World

◆

Say you emerge from your shelter to find that nothing is going to return to normal. Or perhaps a plague has wiped out ninety-nine percent of the world's population. It's a dead earth. What do you do? Living alone is an option, but searching out survivors is a necessity. You really should, as part of your continency plan, organize a post-apocalyptic survival start team. These will be the people you have chosen to help start your new civilization. Include them in the continency plan. If you find yourself alone, or with minimal people, your best bet is to gather survivors and try to begin again. By doing this, you'll show strength that a lot of people will need. Below is a list of things and people, simply phrased, that you will need to think of when planning.

A new civilization area. Find a piece of land, large and fertile that has access to fresh water, or a water supply system you can get up and running. Preplan this, or sit down with your council (Starters) and designate a place.

Find the people. The first people you will need are strong, heathy laborers. Skills are not important with this first batch. Don't waste time finding a doctor, farmer, electrician and so forth. Everyday man is intelligent enough to adapt and learn the acquired skills. Most doctors have never gotten their hands dirty. Laborers are needed for moving equipment, building, and farming.

Survival runs. After you have a minimal group together, plan a survival run. Break off into small groups, sending each group out with a topic of supplies. For example, one group may be clothing, And they will fill one eighteen wheeler with winter, summer, and all purpose clothing. Another group will search out long term food supplies, another…medical.

Books. You will need lots of books on different subjects. Your people have to learn.

Farmers are you next important group of people. Doctors are your last.

Remind your new society that it is the goal to rebuild the world. Bringing new life into the world will be important. However, it would be wise to encourage people to wait. Hold off until stability in supplies and environment is reached.

Hardcore Tips

◆

I compiled a list of tips that are fact based and common sense based. You may find them harsh, sometimes crass, and at moments offensive. But these can and will help you in a time when you could be faced with your own survival.

Always keep your wits about you. Stay smart, stay on top. Never project anything less than confidence to those in your shelter.

Never make a final sweep of your kitchen cabinets for food unless you absolutely need to. If you have a stocked basement shelter, leave the kitchen cupboard food there. If looters should enter your home and see empty cabinets, they will know you didn't carry that food very far. The first place they are going to go for food is your basement and after you.

People will kill for food. They will kill you.

For boiling a gallon of water, a fist size stone, heated in a camp fire can be dropped in a pot of water to bring it to a boil.

Never boast or brag to a neighbor about your shelter or supplies. Unless they are part of your plan, no one should know what you are planning.

Bring only those you trust into your survival plan.

You can drink your own urine.

It is best to leave the family dog out of the shelter. He is another mouth to feed and another source of infection.

If it is necessary, you can eat the family dog. As offensive as it sounds, keep in mind, when hungry, that dog will eat you.

A cat should be kept as close to the shelter as possible but not in, the fecal material in cats spreads disease. But an uncontaminated cat who doesn't have signs of radiation poisoning is an excellent source of food.

It is true that only the strong survive. Weak people should have no place in your new world. If they can't carry the weight, if they emotionally weak, lose them.

Be selective about who you place in your new society, post-apocalyptic group. If you have an acquaintance that you can not stand, or dislike, unless they can be a valuable asset, do yourself a favor and do not include them. The apocalypse is mentally strenuous enough without having a predisposition with someone.

Do not let strangers, or sick people into your shelter.

Do not waste your food and water on dying people.

Establish rules, and leadership right off the bat. This will prevent confusion down the line.

Children will be our most valuable asset for the future. Protect them at all cost. They will be the easy targets of post-apocalyptic survivor syndrom.

Know what your are talking about. Know your facts. Be informed ahead of time.

Survival Review Quiz

◆

It was quite the debatable issue for me on how to end this book. I contemplated the thought of a epilogue, or after word, but what could be said that already wasn't. The entire last section is, to me, the 'wrap up'. Instead of reiterating in a few paragraphs, I decided to end this book with a quiz. Take a moment, after reading, to try your hand at the quiz. Then go to the answer page and see how you did. Good luck.

What does RISK stand for?

Having a backup, then another backup to your plan, plus other details hashed out, is called?

True or false, anyone under the age of fifteen should not be included in survival planning.

True or False, as long as your house is still standing after a nuclear blast, it's safe to stay anywhere in the house.

True or False. Dehydrated foods make the best long-term food because of shelf life and weight.

Over half the people who survived a blast will die from this.

The type of soil best used for making a filtering system is:

True or false, boiling water removes radioactivity.

How long is the standard recommendation for waiting before you leave the shelter?

This is the name of the radiation until that is measured per hour.

What is the dosage of radiation a person can safely be exposed to before showing symptoms?

Name the four types of chemical agents.

Name the three ways you can be infected with Anthrax.

True or false. You should be humane and let anyone, sick or not, into your shelter.

If your home is standing, this is a built in viable means of safe water supply especially in an emergency hit.

What is the number one item people forget to bring into a shelter?

Answers

— ◆ —

Here are the answers to the previous quiz. See how you did, they use the scale below to determine how survival savvy you actually are.

Readiness, Information, Sensibility, Keen Awareness and Actions

The Contingency Plan

False

False

True

Water Borne Illness

Clayey

False

Two Weeks

Roentgens

100R

Nerve, Blood, Choking, Blister

Inhalation, digestion, skin.

False

Water heater

Can opener

How did you score?

Check you score and add them up. See how you did.

13-16   You are a survival champ. You pay attention to detail and mastered a lot throughout this book.

9-12   You did above average, and should be proud, You would do good in a survival situation.

6-8   Below average, but still have some knowledge. You may want to refresh on the sections.

Anything five and below you may want to reconsider re-reading the book. The questions were very basic.

I hope you enjoyed, *Surviving the Apocalypse.* May it bring you the information needed, should you ever be faced with…the apocalypse.

About the Author

◆

Jacqueline Druga-Marchetti is a native of Pittsburgh. She is a full time writer and musician who resides with her four children. Under various genres her published works include: *Siege of the Witches*, *The Peacekeeper*, *The Silent Victor*, A Letter to Sergeant Gillian, *The Nerdfly*, and *BOB*.

Jacqueline Druga-Marchetti welcomes your comments regarding this and any of her other works via email at **info@beyondpassing.com**

References

◆

The Bible, various authors, Tyndale House Publishers, 1996

Various Conversations and web visits: *Centers For Disease Control* (CDC).

Federation of American Scientists. Their website is a vat of knowledge for anyone seeking a more in-depth look at the science of things visit: http://www.fas.org.

Nuclear War Survival Skills, Cresson H. Kearny. An excellent read and most informative. Caroline House, NWS Research Bureau, 1982

Personal and Family Survival, Department of Civil Defense. Publication SM 3-11., 1963

Public Broadcasting System Website, very educational, http://www.pbs.org

The Survivalist's Little Book of Wisdom, David Scott, ICS Books, Inc., 1997